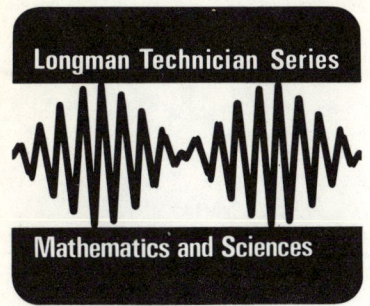

Sector editor:

D.R. Browning, B.Sc., F.R.S.C., A.R.T.C.S.
Principal Lecturer and Head of Chemistry, Bristol Polytechnic

Books already published in this sector of the series:

Fundamentals of chemistry *J.H.J. Peet*
Further studies in chemistry *J.H.J. Peet*
Physical sciences Level 1 *D.R. Browning and I. McKenzie Smith*
Engineering science for technicians Level 1 *D.R. Browning and I. McKenzie Smith*
Safety science for technicians *W.J. Hackett and G.P. Robbins*
Science formulae for TEC courses *D.R. Browning*
Technician mathematics Level 1 *J.O. Bird and A.J.C. May*
Technician mathematics Level 2 *J.O. Bird and A.J.C. May*
Technician mathematics Level 3 *J.O. Bird and A.J.C. May*
Technician mathematics Levels 4 & 5 *J.O. Bird and A.J.C. May*
Mathematics for electrical and telecommunications technicians Level 2
     J.O. Bird and A.J.C. May
Mathematics for electrical technicians Level 3 *J.O. Bird and A.J.C. May*
Mathematics for electrical technicians Levels 4 & 5 *J.O. Bird and A.J.C. May*
Calculus for technicians *J.O. Bird and A.J.C. May*
Statistics for technicians *J.O. Bird and A.J.C. May*
Mathematical formulae for TEC courses *J.O. Bird and A.J.C. May*

Contents

Contents

Preface

This book provides an introduction to chemistry and assumes no prior knowledge of the subject whatsoever. It is intended for students on TEC (Level I), GCE 'O' level, CSE and other elementary courses.

The book includes all the material which is required for the TEC Level I standard unit, although the topics appear in a somewhat different order. We have been very concerned when writing this book to present facts in a logical sequence, and have tried to avoid *forward* references wherever possible. This treatment in no way conflicts with the requirements of the standard unit, which is published primarily as an assessment unit rather than a teaching unit.

We have not confined ourselves rigidly to the standard TEC unit, but have introduced certain other ideas, *e.g.* the periodic table, in order to provide a balanced treatment and to give students some understanding of the considerable amount of factual material that must always be assimilated at this level. By broadening the basis of the book in this way we hope to meet the needs of students working to non-standard units, and also to cater for those on 'O' level and CSE courses. We have certainly not covered the entire 'O' level and CSE syllabi — indeed, this is not our intention — but we do claim to have dealt with all the basic facts and principles that constitute the core of every elementary chemistry syllabus.

Nomenclature in this book is IUPAC, rather than ASE, and units are SI. (A section on SI units appears in Chapter 1.) At the end of each chapter there is a summary of objectives and also several multi-choice and short questions. The questions are provided primarily for consolidation purposes, but they could also prove useful in the setting of phased assessments under the TEC scheme.

We should like to express our thanks to David Browning of Bristol Polytechnic, who has edited this book, and to all those at Longman's University and Further Education Division who have worked on its production.

John Brockington
Peter Stamper
May 1981

Chapter 1

Introduction to chemistry

What is chemistry?

The modern science of chemistry has its origins in the mediaeval occupation of alchemy. This was the haphazard and unsuccessful search for the 'philosopher's stone', which, it was believed, would turn base metals, such as lead, into gold. The popular image of chemistry as a schoolboy pastime concerned with bangs, flashes and unpleasant smells owes much to alchemy, but is far removed from present day reality. The truth is that in recent years chemistry has progressed so far, and extended its frontiers so widely, that it now embraces the whole of the material world. Although it is difficult to summarise such a broad subject in a few words, chemistry can perhaps best be described as the study of substances.

This immediately raises the question, 'What is a substance?' The term is in common use and has various meanings, but in chemistry it is generally understood to mean a single, pure material. Chemistry deals with *all* substances, including those which occur naturally and those which are man-made, and is particularly concerned with the following aspects:

(*a*) the *isolation* of substances, *i.e.* obtaining them in a pure state,
(*b*) the *structure* of substances, *i.e.* the manner in which they are built up,
(*c*) *chemical changes*, *i.e.* changes involving the formation of new substances.

We shall now consider each of these aspects in a little more detail.

Isolation

Pure substances are relatively uncommon. Most things that we encounter in everyday life are *mixtures, i.e.* two or more substances mixed together so

Table 1.1 Percentage, by volume, of the principal gases in the atmosphere

Gas	Percentage
Nitrogen	78.09
Oxygen	20.95
Argon	0.93
Carbon dioxide	0.03

that each retains its original nature. Air, for example, does not consist of a single gaseous substance, but is a mixture of several gases, mainly nitrogen and oxygen. The approximate composition of dry air is shown in Table 1.1. Ordinary air contains, besides these gases, a variable amount of water vapour.

Some other examples of mixtures are shown in Table 1.2.

Table 1.2 The composition of some commonplace materials

Material	Principal substances present
Steel	Iron, carbon
Wine	Water, ethyl alcohol, sugar
Vinegar	Water, acetic acid
Sea water	Water, common salt
Soda water	Water, carbon dioxide

Water, however, is a single substance, and in the form of rainwater is almost pure. It is not absolutely pure, because it contains a small proportion of air. Tapwater is less pure than rainwater because it contains, in addition to air, varying amounts of other substances. This is easily demonstrated by pouring a small quantity of tapwater on to a watch glass and allowing it to dry. A white deposit remains on the glass.

Another almost pure substance that is familiar to all of us is the aluminium of which kettles and saucepans are made. Aluminium which has been exposed to the air is never quite 100 per cent pure because the surface of the metal becomes coated with a very thin layer of a white solid called aluminium oxide. The layer becomes thicker as time goes by and the metal becomes increasingly dull.

The separation of mixtures and the isolation of substances in the pure state is described in Chapters 2 and 7 of this book.

Structure

All substances are built up from tiny particles called *atoms*. Most substances contain only a few different sorts of atoms, although over a hundred kinds are now known to exist. Every substance has its own particular *structure*, which is the term we use to denote the way in which the atoms are arranged together. Some substances, like sodium chloride (common salt) have relatively simple structures, while others, such as those which make up plants and animals, are extremely complicated. Establishing these structures often requires years of patient work, but is a necessary part of chemistry. A knowledge of the structure of a substance helps in understanding its *properties*, such as its behaviour on heating, or when brought into contact with other substances.

We shall study atoms in Chapter 3 of this book, and the structure of substances in Chapters 4, 5 and 8.

Chemical changes

More than anything else, chemistry is concerned with chemical change. Many such changes take place all around us in everyday life, and we tend to take them for granted. One example is the rusting of iron, *i.e.* the conversion of iron into a completely different substance called iron oxide. Another simple example is the burning of methane (North Sea gas) in a gas fire, bunsen burner, etc. to give two other substances, namely carbon dioxide and water vapour. The burning of *any* substance involves chemical change.

All forms of growth and decay take place by chemical change. Plants grow by absorbing water and nutrients from the soil, together with carbon dioxide from the air, and converting them into a substance called cellulose. Animals (including man) grow by converting food into the substances which make up flesh and bone. The rotting of dead plants and animals provide further instances of chemical change.

There are many chemical changes associated with food and drink. Cookery often involves chemistry; for example, when we boil an egg we convert the colourless fluid albumen (*i.e.* the white of the egg) to a jelly-like white solid. The brewing of beer also involves chemical change; in this instance, the conversion of sugary substances in the unfermented liquor into alcohol and carbon dioxide. Other chemical changes occur when food and drink are digested.

It is very difficult to reverse most chemical changes so as to recover the original substances. For instance, when iron has rusted away, we cannot change the iron oxide back into iron except by feeding it into a blast-furnace. It is even more difficult to reconvert carbon dioxide and water into methane, rotten wood into new wood, beer into unfermented liquor, digested food into fresh food, etc.

We must not make the mistake of thinking that a chemical change always takes place when two or more substances are mixed together. An immediate chemical change is relatively uncommon. Sometimes it can be brought about by the use of heat or electricity, but very often the substances concerned remain simply as a mixture and there is no chemical change. No amount of persuasion will make, say, a mixture of sugar and salt undergo a chemical change.

The difference between chemical changes and physical changes

By no means all the changes that we encounter are chemical. There are many which do not involve the formation of new substances, and they are known as *physical changes*. If we move something, or melt it or boil it, we do not as a rule convert it into something else. Thus, the pumping of water from one place to another is an example of physical change. So, too, is the melting of ice and the boiling of water. Most physical changes are readily reversed. We can, for instance, easily pump water back to its original location, reconvert water to ice, or steam to water.

All changes in nature are either chemical or physical, and we must be able to say with certainty which are which. Ease of reversibility, although a good guide, is not entirely reliable, for there are some chemical changes which are easily reversed and some physical changes which are very difficult to undo. Two examples will illustrate this.

i.   The well-known liquid metal, mercury, combines with oxygen in a chemical change to give a red powder called mercury oxide. However, if this red powder is heated, it breaks down again into mercury and oxygen.
ii. The grinding of wheat into flour is properly classified as a physical change because no new substance is formed. However, the reassembly of minute particles of flour to give grains of wheat is a practical impossibility.

The only reliable test of whether a change is chemical or physical is whether or not new substances are formed.

Chemical properties

The chemical changes undergone by a substance are a result of its *chemical properties*. They contrast with *physical properties*, such as melting temperature and boiling temperature. Physical properties can be studied without bringing about any permanent alteration in the nature of the substance concerned; chemical properties always relate to the formation of other substances.

The scope of chemistry

Chemistry enters into any subject in which materials are involved. Engineering, for example, requires a knowledge of fuels, corrosion, electroplating and the treatment of boiler feed water, all of which are chemical topics. Medicine, dentistry, forensic science, metallurgy, agriculture, printing, and even such unlikely subjects as catering and hairdressing, can all benefit from a knowledge of the facts and principles of chemistry.

The following are perhaps the most important areas in which chemistry is applied.

Chemical technologies

Chemistry is essential in the manufacture of most materials. Some of the best-known examples from an almost endless list include petrol, oils, detergents, plastics, paint, rubber, glass, ceramics, fertilisers, pesticides, explosives, drugs, textiles, dyes, food and beverages. Around each of these items has grown a *chemical technology*, concerned with the application of chemistry to the development and production of the material concerned.

Analytical chemistry

Many chemists are employed as analysts, finding out what is present in certain materials (*qualitative analysis*) and how much is present (*quantitative analysis*). Analysis is carried out in laboratories wherever chemistry is practised, and is particularly valuable in the field of research into new substances. In industry, the main use of analytical chemistry lies in controlling the quality of products. All manufactured or processed materials, especially food and drink, must be monitored to ensure that they are of the required standard and do not contain any unpleasant impurities.

Synthetic chemistry

Some of the substances that chemists wish to study occur naturally, perhaps in the ground, the sea or the air. Others can be obtained from plants and animals. But a great many substances are not naturally occurring, and chemists must *synthesise* these, *i.e.* prepare them from materials that are readily available. Most chemical technologies are based on the work of synthetic research chemists.

Biochemistry

Because the cells of living plants and animals are formed and maintained by chemical reactions, a biologist must have a good knowledge of chemistry if he is to understand these processes. Diseases are associated with chemical changes which must be appreciated if the diseases are to be successfully treated by drugs.

The chemistry of biological processes is known as *biochemistry*.

SI units

Chemists, like other scientists, often have the job of measuring physical quantities such as length, volume, time, temperature and electric current. If people in different sciences and in different countries are to work together efficiently, they must use the same *units, i.e.* the same basic quantities of measurement, and the metric system of units was introduced for this purpose in 1875.

In 1960 the metric system was extensively revised by the General Conference on Weights and Measures. The set of units that resulted from this revision is known as the Systeme International d'Unites, and has the cipher SI.

All physical quantities can be divided into *basic quantities* and *derived quantities*. SI recognises seven independent basic quantities; they are shown, together with their units, in Table 1.3. These are, by international agreement, the most *convenient* basic quantities. There is nothing fundamental about the number or, indeed, the choice of these basic quantities, and other lists could conceivably have been adopted.

Derived quantities are defined in terms of basic quantities, and have units which are derived from basic SI units by suitable multiplication or division.

Table 1.3 Basic SI units

Physical quantity	Name of unit	Symbol for unit
Length	Metre	m
Mass	Kilogramme†	kg
Time	Second	s
Electric current	Ampere	A
Temperature	Kelvin	K
Luminous intensity	Candela	cd
Amount of substance	Mole	mol

† The UK spelling is kilogram'

For example, 'velocity' is a derived quantity because it is defined in terms of length and time. The unit of velocity is the metre per second, represented by $m\ s^{-1}$. Many of the more complex derived units, such as those for energy and force, have special names and symbols (see Table 1.4).

So far we have considered only *coherent* SI units, *i.e.* basic units or those derived from basic units without the introduction of numerical factors. In practice, however, they may not be the most convenient ones to use, and other units are obtained from them by the use of multiples and sub-multiples (see Table 1.5). For example, in chemistry it is customary to use the kilojoule, kJ, as a unit of energy, and the cubic decimetre, dm^3, as a unit of volume. (The special name of 'litre' for the cubic decimetre is permissible.)

Only multiples of 3, 6, 9, etc., or sub-multiples of -3, -6, -9, etc., are in common use. Thus, it is conventional to speak of the kilojoule (10^3 J) or

Table 1.4 Derived SI units used in chemistry

Physical quantity	Name of unit	Symbol for unit
Area	Square metre	m^2
Volume	Cubic metre	m^3
Density	Kilogramme per cubic metre	$kg\ m^{-3}$
Energy	Joule	$J\ (kg\ m^2\ s^{-2})$
Force	Newton	$N\ (kg\ m\ s^{-2} = J\ m^{-1})$
Pressure	Pascal, or newton per square metre	$Pa\ (kg\ m^{-1}\ s^{-2} = N\ m^{-2})$
Specific heat capacity	Joule per kilogramme kelvin	$J\ kg^{-1}\ K^{-1}\ (m^2\ s^{-2}\ K^{-1})$
Concentration	Mole per cubic metre	$mol\ m^{-3}$
Electric charge	Coulomb	$C\ (A\ s)$
Electric potential difference	Volt	$V\ (kg\ m^2\ s^{-3}\ A^{-1} = J\ A^{-1}\ s^{-1})$
Electric resistance	Ohm	$\Omega\ (kg\ m^2\ s^{-3}\ A^{-2} = V\ A^{-1})$

Table 1.5 Prefixes for multiples and sub-multiples

Multiplying factor	Prefix	Symbol	Multiplying factor	Prefix	Symbol
10^6	Mega	M	10^{-2}	Centi	c
10^3	Kilo	k	10^{-3}	Milli	m
10^2	Hecto	h	10^{-6}	Micro	μ
10	Deca	da	10^{-9}	Nano	n
10^{-1}	Deci	d	10^{-12}	Pico	p

the megajoule (10^6 J) but not, say, the hectojoule (10^2 J). The centimetre (10^{-2} m) and the decimetre (10^{-1} m) are special cases.

A few non-SI units of pressure are still encountered in chemistry. They are shown in Table 1.6, together with their definitions in terms of SI units.

Table 1.6 Non-SI units of pressure

Name of unit	Symbol for unit	Definition of unit
Atmosphere	atm	101 325 Pa†
Millimetre of mercury	mmHg	$13.5951 \times 9.806\ 65$ Pa†

†For most purposes,˙1 atm = 760 mmHg $\approx$ 100 kPa

Summary

At the conclusion of this chapter, you should be able to:
1. define a 'mixture' as two or more substances mixed together so that each retains its original nature,
2. state that air is a mixture of pure substances,
3. state the main components of air and their approximate proportions,
4. recognise that all substances are built up from atoms,
5. classify changes as physical or chemical and give examples of each,
6. state the basic SI units of measurement.

Questions

Which of the operations in questions 1—8 involve chemical change and which involve physical change?
1. Moving a chair.
2. Chopping up a chair.
3. Burning a chair.
4. Forming salt crystals by leaving sea water exposed to the sun.
5. Smoking a cigarette.
6. Winding up a clockwork toy.
7. Letting off a firework.
8. Grilling a steak.

Chapter 2

Substances

Before a substance can be studied properly it must be purified. This is because the presence of impurities, even in small amounts, can drastically alter the properties of a substance. We therefore need to learn how mixtures can be separated so as to give, in a pure state, all the substances which they contain. We also need some means of knowing whether or not substances are pure. These are among the topics that we shall be discussing in the first part of this chapter. Afterwards, we shall see that there are two basic types of substances, namely elements and compounds.

Isolation of substances

There are various methods that can be used for the separation of mixtures, and three of them, namely decantation, filtration and sublimation, are described below. The techniques of evaporation, crystallisation and distillation, which are employed for the separation of solutions, are considered in Chapter 8.

Decantation

A mixture of a liquid, such as water, and a coarsely divided solid, such as sand, can often be separated merely by *decanting* the supernatant liquid, *i.e.* by pouring it off carefully, once the solid has settled to the bottom of the containing vessel. Care must be taken at the end to ensure that none of the solid is lost with the last few drops of liquid. Alternatively, the liquid can be drawn off by means of a teat pipette (see Fig. 2.1).

In order to obtain a good separation the following procedure should be adopted.

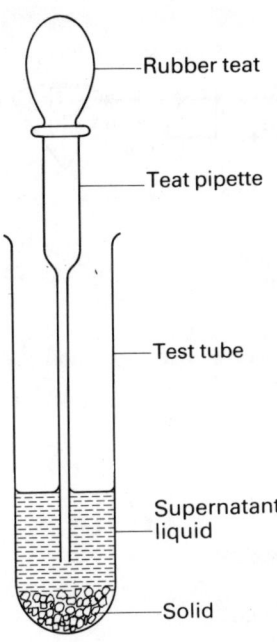

Fig. 2.1 Use of a teat pipette

i.   Squeeze the teat *before* inserting the pipette in the test tube.
ii.   Ensure that the nozzle of the pipette is positioned slightly above the
      solid.
iii. Release the teat carefully, so as not to disturb the solid.

   Decantation does not give a perfect separation; in this example, the sand
is left in a wet condition. Provided that the liquid is volatile, *i.e.* provided it
evaporates at a reasonable rate, this does not matter. Wet sand, for instance,
can easily be dried in an oven.

Filtration

Decantation is unsatisfactory whenever the solid is finely divided or gelatinous,
i.e. jelly-like. In such cases the solid does not fall rapidly to the bottom of the
vessel, but floats in the liquid. Mixtures of this sort can usually be separated
by *filtration*. This is the term applied to the separation of a solid from a
liquid by a sieving action. The apparatus consists basically of a filter funnel
(see Fig. 2.2(a)), a filter paper (see Fig. 2.2(b)) and a beaker or flask to receive
the *filtrate, i.e.* the liquid that has passed through the paper (see Fig. 2.2(c)).
The filter paper is effectively a sieve, which allows the liquid to pass through
its pores, but not the particles of solid.

   Before it can be used, a filter paper must first be folded in half, then in
half again, and opened out so as to make a cone (see Fig. 2.2(b)). It is then

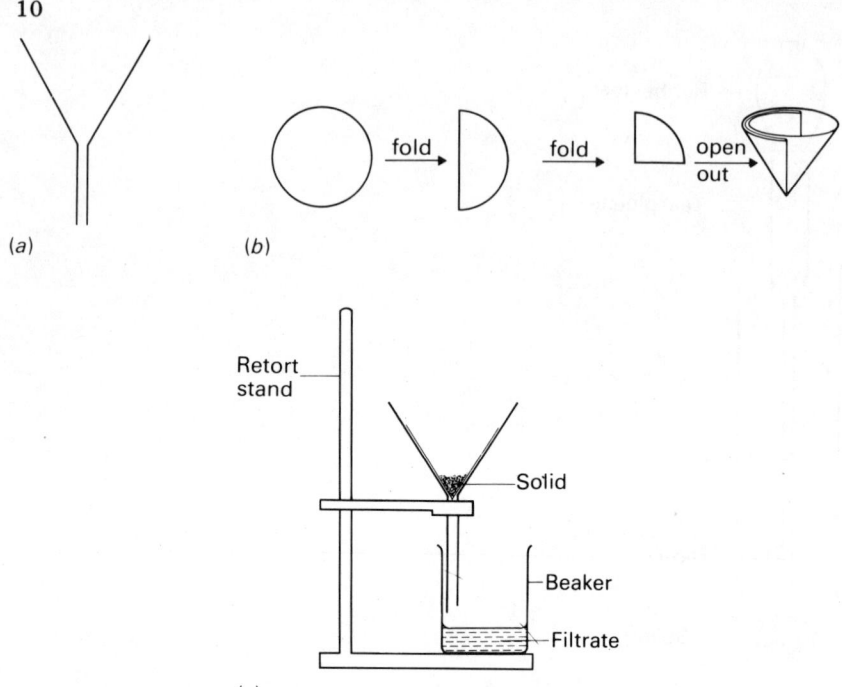

(a) (b)

(c)

Fig. 2.2 Filtration equipment

inserted in a funnel which is supported in a vertical position with its stem touching the side of the beaker. The paper is pressed down and wetted so that its upper portion adheres to the funnel, and the mixture to be separated is poured in. Liquid drops through the funnel into the beaker, whilst solid remains in the paper. When filtration is complete, the filter paper can be removed from the funnel and opened out to dry on a clock glass, *i.e.* a shallow, circular glass dish. Obviously, the paper and its contents will dry more quickly in the oven than at room temperature.

Notes on filtration

1. Filter funnels are available in a wide range of sizes, and care should be taken to select one that will be no more than one-third full of solid at the end of the operation.
2. Filter papers, likewise, are available in a range of sizes to fit all funnels. It is important to choose one of the correct size, so that the folded paper extends to within 1–2 cm of the top of the funnel, but never closer than 1 cm.
3. The filter paper should be wetted before filtration with the liquid which is present in the mixture to be separated. This will not always be water.
4. A funnel must never be overfilled. If liquid is allowed to rise over the top of the filter paper it will inevitably carry with it some of the solid. This

will penetrate the space between the paper and the glass, and eventually contaminate the filtrate.

Filtration has an advantage over decantation in that fine particles of solid are less likely to be lost. There is, however, a disadvantage, in that filtration can be rather slow. The trouble can be minimised by selecting a filter funnel with a long, narrow stem. (A length of 15—20 cm is recommended.) The reason is that during filtration a column of liquid builds up in a narrow stem, and the weight of this liquid helps to draw further liquid through the paper.

Sublimation

Most solid substances, when they are heated, melt to form a liquid. Ice, for example, melts to give water. The liquid, on further heating, gives rise to a vapour. There are some solids, however, which on heating pass straight to a vapour, without going through a liquid stage. The vapour, when condensed, reverts to a solid. A solid substance of this kind is said to 'sublime' on heating. If it is mixed with an involatile substance, *i.e.* one which does not vaporise, it can be purified by *sublimation*. Essentially, the impure solid is heated, so that it sublimes, and the vapour is condensed to a solid on a cool surface, *e.g.* the inside of a filter funnel (see Fig. 2.3). A water-cooled condenser, such as is used for distillation, cannot be employed, because it would soon become blocked by *sublimate, i.e.* the solid that has sublimed.

The technique is readily demonstrated with a mixture of iodine, which sublimes, and sand, which does not. (*Note*: **Iodine vapour is toxic, and this experiment must be carried out in a fume cupboard.**) The mixture is placed in an evaporating basin, which rests on a tripod and gauze. A filter paper is perforated in several places with a pencil, and placed over the evaporating

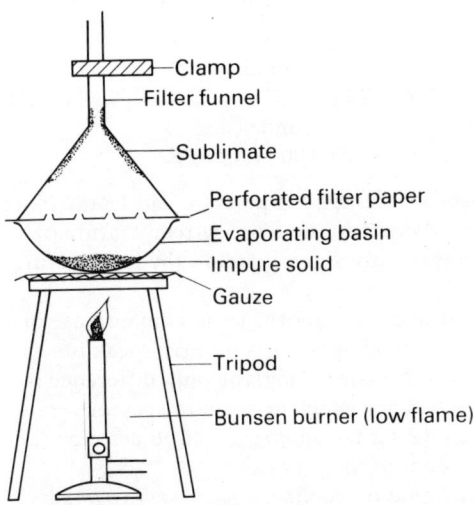

Fig. 2.3 Sublimation apparatus

basin so that the jagged edges of the holes face upwards. The paper is then covered with an inverted filter funnel, of diameter slightly less than that of the basin. All the holes in the paper should be covered by the funnel.

On heating the basin, iodine sublimes through the holes in the paper and collects partly on the top of the paper and partly on the walls of the funnel. The funnel remains relatively cool, because the filter paper acts as a heat insulator. Nevertheless, the basin must be heated gently, for otherwise much of the vapour escapes through the stem of the funnel. Draughts, likewise, can lead to the loss of vapour, and should be minimised by the use of screens.

The simple apparatus described here is clearly unsuitable for the sublimation of valuable substances. If loss of material is to be avoided altogether, a specially designed apparatus with an efficient condensing system must be used.

The three states of matter

We often classify substances as solids, liquids or gases, depending on their state at room temperature and atmospheric pressure, but in fact virtually every substance can exist in all three states according to the conditions. Consider water, for example. It can exist in the solid form, 'ice'. On warming, ice changes abruptly at a fixed temperature called its *melting temperature* to the liquid form, 'water', and on further heating this is converted, at the *boiling temperature*, to the gaseous form, 'steam'. There is no chemical change when a solid passes to a liquid or when a liquid passes to a gas, and ice, water and steam are simply different forms of the same substance.

Melting and boiling are examples of physical change and, as we might expect, they are easily reversed. Steam can be *condensed* to water merely by cooling, and on further cooling water *freezes* to ice. The changes may be summarised thus:

	melting		boiling	
ICE	$\longleftrightarrow$	WATER	$\longleftrightarrow$	STEAM
	freezing		condensing	
	At 0 °C (273 K)		At 100 °C (373 K)	

Both temperatures relate to changes at atmospheric pressure, *i.e.* 1 atm. Note that the melting temperature of ice is equal to the freezing temperature of water; likewise, the boiling temperature of water is equal to the temperature at which steam condenses.

Certain other terms, besides those quoted above, are in common use in this context. For instance, instead of referring to a gas we may speak of a 'vapour'. The two terms mean almost the same thing; the only difference is that the term 'gas' is applied to a substance which exists in the gaseous state under ordinary conditions, whereas the term 'vapour' is used to describe the gas into which a liquid (or solid) is converted by heat.

While a vapour condenses to a liquid on cooling, a gas is said to *liquefy*. Thus, we speak of the 'condensation of steam', but the 'liquefaction of nitrogen', although essentially both processes are the same. There is a similar distinction between freezing and solidification. A substance which is a liquid at room temperature freezes to a solid on cooling, whereas a substance which normally exists as a solid is said to *solidify* when the liquid form is cooled. Thus, water 'freezes', but molten iron 'solidifies'.

By contrast, there is a fundamental difference between evaporation and boiling, for whereas *evaporation* from the surface of a liquid occurs to some extent at all temperatures, *boiling* takes place only at the boiling temperature. Boiling entails evaporation not only from the surface but also from the body of the liquid. Consequently, when a liquid boils, bubbles of vapour form beneath the surface.

Although a change of state is usually brought about by an alteration of temperature, it can also be caused by a change of *pressure*. This applies particularly to the change from the liquid to the gas state (or vice versa). Water, for example, can be converted to steam not only by raising the temperature but also by lowering the pressure. It is well known that at high altitudes, where the pressure is less than ordinary atmospheric pressure, water boils at a temperature below $100\,^{\circ}C$ (373 K); so much so that it may not be possible to make a good cup of tea on top of a mountain! It is perhaps less well known that water will even boil at room temperature if the pressure is reduced sufficiently. All liquids can be vaporised by a reduction of pressure.

The converse is also true. Steam, for instance, can be changed into water by increasing the pressure, even though the temperature is over $100\,^{\circ}C$ (373 K). All gases can be liquefied by an increase of pressure, provided that the temperature is not too high.

The effect of pressure on the change from the solid to the liquid state, or vice versa, is insignificant for most practical purposes.

The principles described above are of general application, although there are a few exceptions. Some substances, for example, exist only as solids because they decompose on heating before a melting temperature is reached. This applies to sugar, which chars on heating, and also to calcium carbonate (limestone), which breaks down below its melting temperature into calcium oxide (quicklime) and carbon dioxide. Likewise at atmospheric pressure some liquids decompose on heating before a boiling temperature is reached. For such liquids it is customary to quote a boiling temperature at a specified reduced pressure. Finally, as we have seen above, there are a few substances which do not exist in the liquid state at atmospheric pressure, but sublime from the solid to the gaseous condition on heating.

Summary
1. Pure substances exist as solids, liquids or gases.
2. The physical state of a pure substance depends on the prevailing conditions, namely temperature and pressure.
3. The principal condition that controls the change from solid to liquid (or vice versa) is temperature; pressure has relatively little effect.
4. Both temperature and pressure play an important part in the change from liquid to gas (or vice versa).

Melting temperatures and boiling temperatures

Every pure solid has its characteristic melting temperature, formerly called 'melting point', which is very sharp. Ice, for example, when warmed from, say, $-10\,^{\circ}C$ (263 K), shows no sign of melting until a temperature of $0\,^{\circ}C$ (273 K) is reached. Even at $-1\,^{\circ}C$ (272 K), the ice is entirely solid; but at $0\,^{\circ}C$ it melts completely. Granted, the ice may take a long while to melt, and heat must be supplied to help the melting process, but the point we must appreciate here is that **all the ice melts at $0\,^{\circ}C$.** In a similar manner, naphthalene (used for making mothballs) melts sharply at $80\,^{\circ}C$ (353 K), and sodium chloride (common salt) at $801\,^{\circ}C$ (1074 K). For all pure solids these temperatures are quoted, usually in tabular form, in textbooks of practical chemistry.

In the same way that every solid has its own melting temperature, so every liquid has its own, sharp boiling temperature or 'boiling point'. Pure water, for instance, at atmospheric pressure, does not boil at all at $99\,^{\circ}C$ (372 K), but boils away completely at its boiling temperature of $100\,^{\circ}C$ (373 K). Ethanol (ethyl alcohol) has a boiling temperature of $78\,^{\circ}C$ (351 K); ethanoic acid (acetic acid) boils at $118^{\circ}C$ (391 K), and so on. The values for all common liquids can be referred to in boiling temperature tables. Like the melting temperatures of solids, they relate to atmospheric pressure unless otherwise stated.

Determination of melting temperatures and boiling temperatures

The melting temperatures of solids which melt below $360\,^{\circ}C$ (633 K) are determined by means of a *melting point apparatus* (see Fig. 2.4).

The apparatus consists basically of an electrically heated block in which are drilled a number of holes. A large hole is provided to hold a thermometer, while the smaller ones are for melting point tubes. The equipment is usually provided with three controls, namely an on/off switch, a knob to govern the rate of heating, and a booster switch for rapid heating to a high temperature.

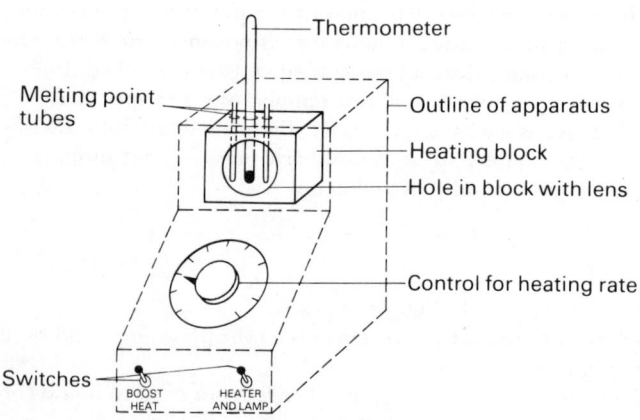

Fig. 2.4 Electrical melting point apparatus

The solid under examination is first powdered, and a small amount of it is introduced into a melting point tube, which is essentially a very narrow, thin-walled glass tube sealed at one end. The tube is inserted, together with a suitable thermometer, into the heating block. When the apparatus is switched on the contents of the tube become illuminated and are visible, slightly magnified, through a lens. The control knob is rotated clockwise so as to give a moderate rate of heating, and the contents of the tube are watched closely. The melting temperature is recorded when the solid is completely molten.

Unless the heating is very gentle in the vicinity of the melting temperature, the method as just described will give too high a value and is suitable only for a first estimate. The determination should be repeated, once the apparatus has cooled somewhat, with a fresh sample of solid in a new melting point tube. This time the tube is heated rapidly to about ten degrees below the approximate melting temperature, and then very carefully, at no more than two degrees per minute, until the solid has just melted.

The boiling temperatures of liquids which boil below 360 °C (633 K) can be found by distillation (see p. 88). Care must be taken to distil the liquid at a moderate rate. If the heating is insufficient, the thermometer bulb does not become totally surrounded by hot vapour and registers a low temperature. On the other hand, if the bunsen burner is turned on full, heat which is radiated and convected from the red hot gauze to the outside of the still-head causes the thermometer to register a temperature which is too high.

Uses of melting temperatures and boiling temperatures

Melting temperatures and boiling temperatures have two uses in chemistry.

As criteria of purity A pure solid, as we have seen, has a sharp melting temperature. Pure naphthalene, for example, melts entirely at 80 °C (353 K). An impure solid, by contrast, melts over a range of temperature which is *below* that of the pure substance. Impure naphthalene, for example, might start to melt at, say, 71 °C (344 K) and become completely molten at, say, 77 °C (350 K). That being the case, we quote a melting range of 71−77 °C (344−350 K). A sharp melting temperature is therefore a criterion of purity for a solid substance. In other words, it is a test of purity; it tells us whether or not a substance is pure. A melting range tells us that a substance is impure; but does not provide a useful guide as to how much impurity is present.

A pure liquid, likewise, has a sharp boiling temperature, but an impure liquid has not. Instead, it boils over a range of temperature known as a *distillation range*. This may be above or below the boiling temperature of the pure substance, depending on the nature of the impurity. Thus, in the same way that a sharp melting temperature is a criterion of purity for a solid, so a sharp boiling temperature is a criterion of purity for a liquid.

The separation of a mixture is always followed by a determination of melting temperature or boiling temperature to see whether complete purity has been achieved. If it has not, further purification is carried out until a sharp melting temperature or boiling temperature is observed.

In the identification of substances The melting temperature and boiling temperature of a substance help to characterise it, *i.e.* to distinguish it from other substances. Water, for example, is characterised by its boiling temperature of 100 °C (373 K) at atmospheric pressure. Few other substances share this characteristic; therefore, if an 'unknown' liquid prepared in the laboratory boils at exactly 100 °C it is probably water.

Thus, by determining the melting temperature or boiling temperature of a substance, and by comparing the value with those quoted in textbooks, we can make a tentative identification of the substance. Subsequent work must be carried out to confirm the identity, because certain substances have melting temperatures or boiling temperatures which lie very close together.

Elements and compounds

Investigation shows that by suitable techniques, frequently drastic, most substances can be decomposed into simpler substances. For example, the red powder commonly called mercury oxide decomposes when heated in a test tube. **The experiment should not be carried out by students because mercury vapour is poisonous, but it can be performed, preferably in a fume cupboard, as a demonstration.** About 2 g of mercury oxide is weighed in a small test tube, which is then heated gently in a bunsen flame. A colour change is soon observed. From time to time a glowing wooden splint is introduced into the mouth of the tube, and is seen to relight. When no further change is apparent, the tube is cooled and reweighed. Finally the grey residue is examined by rubbing it with a glass rod.

What happens on heating is that the mercury oxide decomposes into mercury and oxygen. The oxygen, which is a gas, escapes into the atmosphere, and since wood burns very readily in oxygen the gas will rekindle a glowing splint. The loss in weight of the tube, which should have been noticed, provides further evidence for the escape of oxygen. Mercury is far less volatile than oxygen and mostly remains in the tube as a grey deposit, which forms globules of liquid metal when rubbed with a glass rod.

It is possible to make mercury recombine with oxygen by heating it in air just below its boiling temperature, but for health reasons this experiment should not be performed in student laboratories. Mercury oxide, indistinguishable from the original compound, is reformed.

Few substances undergo thermal decomposition (*i.e.* decomposition by heat) as easily as mercury oxide, and many require temperatures that cannot be achieved with a bunsen burner. Some substances can be decomposed by means of electricity. Water, for example, does not decompose on heating until above 1000 °C (1273 K), but if an electric current is passed through it at room temperature it will be found to decompose into two gases, hydrogen and oxygen (see p. 189). The decomposition of water, like that of mercury oxide, can be reversed. Hydrogen gas from a cylinder can be burnt in air, and if the flame is directed on to a cold surface, such as the botton of an evaporating basin containing cold water, a film of condensation appears immediately. This shows that, on burning, hydrogen combines with oxygen of the atmosphere to give water.

For all practical purposes sodium chloride (common salt) is completely resistant to thermal decomposition. It melts (at 801 $^{\circ}$C) and boils (at 1467 $^{\circ}$C) unchanged. However, in the molten state it can be decomposed by electricity into sodium, which is a soft, silvery-white metal, and chlorine, which is a green gas. If sodium is plunged into a gas jar of chlorine at room temperature it catches fire and burns of its own accord as the sodium and chlorine recombine to give sodium chloride.

There are other substances which are affected by neither heat nor electricity but which can be decomposed in other ways. Carbon dioxide, for instance, is a gas which is virtually unaffected by heat and electricity. However, it can be demonstrated that when burning magnesium is introduced, black specks of carbon are formed as the hot magnesium brings about decomposition into carbon and oxygen. When carbon burns in air it recombines with oxygen to give carbon dioxide, with properties identical to those of the starting material. It can be shown, for instance, that the combustion product, like the original gas, produces a 'milkiness' when bubbled through lime water.

But there are other substances which, no matter how hard we try, we cannot decompose into simpler substances. Mercury, oxygen, hydrogen, sodium, chlorine and carbon, for example; none of these can be decomposed by heat, electricity or any other means, and we refer to them as *elements*, to distinguish them from the *compounds* that are formed when they react together. Mercury oxide, water, sodium chloride and carbon dioxide are all examples of compounds.

Naturally occurring substances may be either elements or compounds. In the atmosphere, for example, there is oxygen, nitrogen and argon, which are elements; and carbon dioxide and water vapour, which are compounds.

To summarise, elements combine together to give compounds, and compounds can be decomposed into their elements. The combination of elements and the decomposition of compounds are examples of chemical change.

Summary

At the conclusion of this chapter, you should be able to:
1. separate mixtures of pure substances by decantation, filtration and sublimation,
2. state that pure substances exist as solids, liquids or gases,
3. state that the physical state of a pure substance depends on the physical conditions,
4. recognise that a substance can change its state by alteration of temperature or pressure,
5. state that pure substances have characteristic melting and boiling temperatures,
6. recognise that pure substances are either elements or compounds,
7. distinguish between elements, compounds and mixtures.

Questions

In questions 1–5 select the most appropriate answer, labelled A, B, C or D.

1. Decantation can be used to separate a solid from a liquid whenever the solid;
   A   is very finely divided,
   B   is gelatinous,
   C   is coarse,
   D   floats in the liquid.

2. The rate of filtration can be increased by;
   A   using a short stemmed funnel,
   B   using a long stemmed funnel,
   C   using a very fine filter paper,
   D   piercing a hole in the bottom of the paper.

3. A certain compound, in the pure state, has a melting temperature of 122 °C (395 K). The impure compound could possibly melt at;
   A   120 °C,
   B   124 °C,
   C   117–120 °C,
   D   124–127 °C.

4. Three compounds, called urea, benzamide and phthalic anhydride, all melt at approximately 130 °C (403 K). Mixtures of these compounds with an unknown solid X melt as follows:

X + urea	130 °C
X + benzamide	112–118 °C
X + phthalic anhydride	120–125 °C

   X is therefore;
   A   urea,
   B   benzamide,
   C   phthalic anhydride,
   D   none of these substances.

5. Liquid water can most easily be vaporised by;
   A   raising the pressure,
   B   lowering the pressure,
   C   raising both the pressure and the temperature,
   D   lowering the pressure and raising the temperature.

6. Classify each of the following materials as an element, compound or mixture.
   (a) Oxygen   (f) Sodium chloride
   (b) Water   (g) Carbon dioxide
   (c) Air   (h) Beer
   (d) Iron   (i) Chlorine
   (e) Carbon   (j) Cement

Chapter 3

Atoms

Fundamental particles

All substances, including both elements and compounds, consist of tiny particles called *atoms*. A knowledge of atomic structure is so important to an understanding of the rest of chemistry that we must discuss it at this stage.

The idea that atoms are the 'building blocks' of all matter was first proposed by John Dalton in 1808 in his 'atomic theory of matter'. Dalton thought that atoms could not be divided or destroyed, but in these beliefs he was incorrect. Atoms are assemblies of even smaller particles called *protons, neutrons* and *electrons*. They are known collectively as *fundamental particles* or *sub-atomic particles*. These particles are tightly bound together, and atoms remain intact, or substantially intact, during chemical changes.

There are many different types of atoms. They are all spherical in shape and are extremely small, with radii ranging from 0.037 nm to 0.27 nm. An atom is so small that even if its size were increased a hundred million times it would still be only as big as a pea. (If a pea were scaled up to this extent it would be almost the same size as the earth itself.)

At the centre of every atom there is a tiny *nucleus*, which contains the protons and neutrons. For the smaller atoms the number of neutrons is approximately equal to the number of protons, but for others the neutron content is greater than the proton content. The total number of protons and neutrons in the nucleus is known as the *mass number*, symbol A, of the atom.

The electrons of an atom are in movement around the nucleus, but at different distances from it. Electrons whose distances from the nucleus are approximately equal form a collection that we call a *shell*. Shells are designated

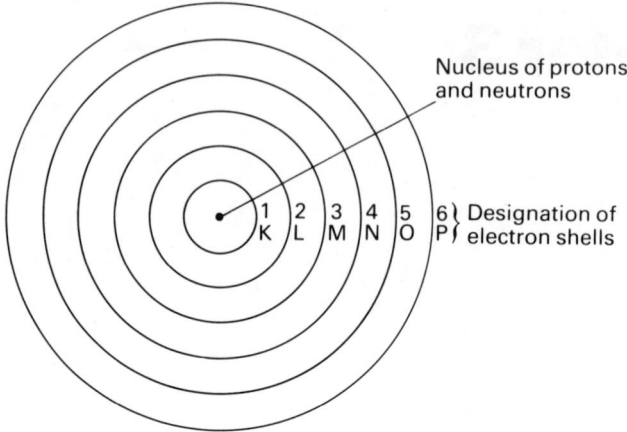

Fig. 3.1 The structure of the atom

either by the numbers 1, 2, 3, 4, 5 and 6, or by the letters K, L, M, N, O and P (see Fig. 3.1). For any atom, the total number of electrons in the shells is equal to the number of protons in the nucleus.

   The electron shells extend relatively far from the nucleus; so far that the size of the nucleus is almost insignificant compared with that of the atom as a whole. To make this clear, suppose that an atom were scaled up so that its nucleus had a radius of 1 centimetre. Then on this scale the radius of the complete atom would be about 100 metres!

   The atoms of any given element are all alike, in the sense that they possess the same number of protons in the nucleus and, therefore, the same number of electrons in the shells. This number is known as the *atomic number* (Z) of the element. Every element has its own particular atomic number, which serves to distinguish it from other elements. Hydrogen, for example, has an atomic number of 1. There is thus one proton in the nucleus of every hydrogen atom, and one electron outside the nucleus. Oxygen, with an atomic number of 8, always has eight protons in the atomic nucleus and eight electrons outside it.

   Although, for any given element, the number of protons in the nucleus is fixed, the number of neutrons can vary within certain limits. This is the reason why most elements can exist in two or more forms, known as *isotopes*. For example, there are three kinds of hydrogen atom. Each has one proton in the atomic nucleus, and one electron in the first shell, but they vary in their neutron content as shown in Fig. 3.2. They are atoms of the three 'isotopes of hydrogen', called, respectively, hydrogen-1, hydrogen-2 and hydrogen-3. The numerical suffixes denote the mass numbers of the atoms. The isotopes of hydrogen also have trivial names (see Fig. 3.2) as well as systematic names, but are unique in this respect. Ordinary hydrogen, as made for example by the electrical decomposition of water, contains more than 99.9 per cent of hydrogen-1, with mere traces of hydrogen-2 and hydrogen-3.

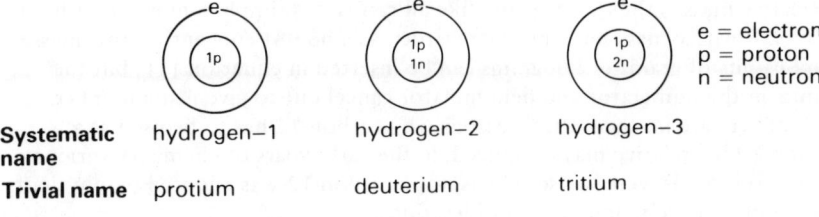

Systematic name	hydrogen–1	hydrogen–2	hydrogen–3
Trivial name	protium	deuterium	tritium

e = electron
p = proton
n = neutron

Fig. 3.2 Isotopes of hydrogen

Almost all the naturally occurring elements consist of a mixture of isotopes. As another example, consider carbon (see Fig. 3.3). Naturally occurring carbon consists of approximately 99 per cent of the isotope carbon-12, and 1 per cent of carbon-13.

carbon–12   carbon–13

Fig. 3.3 Isotopes of carbon

The various isotopes of an element have identical chemical properties. They differ only in certain of their physical properties, such as melting temperature, boiling temperature and density. (*Density* is the mass of a substance divided by its volume.)

Masses and charges of fundamental particles

Mass and relative mass

The amount of matter in a body is known as its *mass*, and mass, in all scientific work, is quoted in SI units of grams or kilograms. The mass of a small particle, like that of a large body, can be expressed in grams; for example, we can say that the mass of a proton is 1.67×10^{-24} g, *i.e.* 0.000 000 000 000 000 000 000 001 67 g. However, the use of high negative indices is inconvenient, and to avoid this it has become the custom to quote the *relative mass* of a small particle rather than its actual mass. 'Relative mass' indicates how the mass of the particle is related to that of an atom of the isotope carbon-12. More precisely, it is defined as the number of times the mass of the particle is greater than *one-twelfth* the mass of a carbon-12 atom, *i.e.*

$$\text{relative mass of a particle} = \frac{\text{mass of the particle}}{\text{one-twelfth the mass of a carbon-12 atom}}$$

[1]

Relative mass is thus a ratio, and like all ratios it is simply a number without units. Relative mass must not be regarded as a non-SI concept. Actual masses in SI units of grams or kilograms can be inserted in equation [1], but the units in the numerator and denominator cancel out to give just a number.

There are no theoretical reasons why carbon-12 has to be used as a standard for relative masses; indeed, in the early years of chemistry various other standards were in use. The isotope carbon-12 was adopted by physicists in 1960 on purely practical considerations.

Relative masses relate to one-twelfth the mass of a carbon-12 atom, and not to the total mass of a carbon-12 atom, so as to give values that are both convenient and traditional. On this basis, the relative mass of a proton is approximately 1, as is that of a neutron. If we decided to relate relative masses to the *total* mass of a carbon-12 atom, then the relative mass of a proton or a neutron would be 1/12, which is not convenient. Moreover, the relative masses of these two particles have always been taken as 1, and not 1/12.

The relative mass of an electron is usually quoted as 1/1840. This is extremely small, and for most chemical purposes can be neglected altogether.

Electrical charge

Whenever you have combed your hair on a dry day, you have probably observed some rather curious effects. Your hair crackles as the comb passes through it, and if it is dark you can see tiny sparks passing between your hair and the comb. You may also have found that, after combing, your hair can be made to stand up straight when the comb is brought near to it.

Why is this? A clue is provided by the sparks that are sometimes noticed. They are sparks of electricity. When a comb is rubbed against your hair, both the comb and your hair become *electrically charged, i.e.* they collect electricity. However, they collect different sorts of electricity, and are said to be 'oppositely charged'. That is why your hair and the comb attract each other. Two oppositely charged articles always attract each other when they are brought close together, but two similarly charged articles repel each other.

The electricity on an isolated comb, or on your hair, does not move and is known as *static electricity*. But when the comb touches your hair electricity flows from one to the other as an *electric current*. The sparks that you can see are due to electricity jumping across small gaps between your hair and the comb, and the crackling sound is caused by the sparks as they pass through the air.

The neutrons in the nucleus of an atom carry no charge. In other words they are electrically neutral; hence their name. Protons and electrons, however, possess electrical charges. They are oppositely charged, and we say that a proton is *positively charged* and an electron is *negatively charged*. Both the proton and the electron carry the same *amount* of charge, *i.e.* the same quantity of electricity.

Electrical charge is measured in coulombs (C), and we can express the charge on a proton or an electron as 1.602×10^{-19} C. Because of the high negative index, however, this is just as inconvenient as expressing the masses of these particles in grams or kilograms. To overcome this problem, we usually quote the *relative charge* on a particle, in much the same way that we quote

Table 3.1 Properties of fundamental particles

Particle	Relative mass	Charge relative to proton
Proton	1	+1
Neutron	1	0
Electron	1/1840	−1

its relative mass. 'Relative charge' is the charge related to that of a proton. We write the charge of a proton as +1; in which case the relative charge of an electron is −1.

The terminals labelled + or − on dry batteries and car batteries can be understood in terms of protons and electrons. A terminal which is positively charged has more protons than electrons, while one which is negatively charged carries more electrons than protons.

The relative masses and charges of fundamental particles are summarised in Table 3.1.

The mass of an atom

Because electrons have virtually no mass, the total mass of an atom is concentrated almost entirely in its nucleus.

As we have seen, both the proton and the neutron have a relative mass of approximately 1. Consequently, the total number of protons and neutrons in an atom is approximately equal to its relative mass; hence the use of the term 'mass number' to express both the relative *mass* of an atom and its *number* of protons and neutrons.

The electricity of an atom

The nucleus

Earlier in this chapter we have seen that:

i.   an atomic nucleus contains one or more protons,
ii. a proton has a relative charge of +1,
iii. the number of protons varies from one element to another, and is known as the atomic number of the element,
iv. an atomic nucleus is very small compared with the size of the atom as a whole.

It follows that an atomic nucleus must be positively charged, with a relative charge that is equal to the number of protons it contains. For example, hydrogen (atomic number 1) has a nuclear charge of +1. Carbon (atomic number 6) has a nuclear charge of +6.

Furthermore, because the atomic nucleus is so small, it carries a very strong positive charge. This was established by Lord Rutherford as long ago as 1909. In one of his experiments Rutherford aimed a stream of fast moving α-particles, *i.e.* nuclei of helium atoms, at a very thin piece of gold foil. He

24

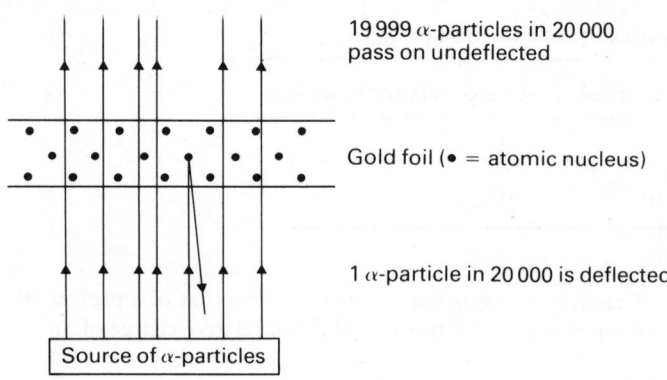

19 999 α-particles in 20 000
pass on undeflected

Gold foil (• = atomic nucleus)

1 α-particle in 20 000 is deflected

Source of α-particles

Fig. 3.4 The bombardment of gold foil with α-particles

found that almost all the α-particles penetrated the foil, only about one in every 20 000 particles being deflected through an angle of 90° or more.

Since α-particles are positively charged, and can change direction only as a result of collision with other positively charged bodies, this experiment provides evidence that the gold foil contains very small, localised regions of positive charge and relatively heavy mass.

Students often ask how it is possible for protons to be packed so closely together in the nucleus, in view of the strong electrical repulsion that might be expected to exist between them. There is, as yet, no satisfactory explanation, although various ideas have been put forward. It has been suggested that the neutrons serve to insulate the protons from one another; also, that when protons are extremely close together the repulsion between them changes to an attraction. The latest theory is that the nucleus may contain small particles called *gluons* that serve to 'glue' the protons together.

Electrons

We have seen that electrons exist around the nucleus in a series of shells (see Fig. 3.1). Because the shells get progressively larger with increasing distance from the nucleus, they can hold more and more electrons. In theory, the number of electrons that any shell can hold is given by the formula $2n^2$, where n is the number of the shell. Thus, the first shell can hold two electrons, the second can hold eight, and so on, as can be seen from the following table:

shell (n)	1	2	3	4	5	6
n^2	1	4	9	16	25	36
number of electrons in shell ($2n^2$)	2	8	18	32	50	72

In practice the higher shells, 5 and 6, are never completely filled.

Electrons are retained in an atom by the electrical attraction exerted by the nucleus. They occupy shells as close to the nucleus as possible. For example,

Table 3.2 Electronic configurations of atoms

Element	Atomic number (Z)	Electronic configuration			
		1st shell	2nd shell	3rd shell	4th shell
Hydrogen	1	1			
Helium	2	2			
Lithium	3	2	1		
Beryllium	4	2	2		
Boron	5	2	3		
Carbon	6	2	4		
Nitrogen	7	2	5		
Oxygen	8	2	6		
Fluorine	9	2	7		
Neon	10	2	8		
Sodium	11	2	8	1	
Magnesium	12	2	8	2	
Aluminium	13	2	8	3	
Silicon	14	2	8	4	
Phosphorus	15	2	8	5	
Sulphur	16	2	8	6	
Chlorine	17	2	8	7	
Argon	18	2	8	8	
Potassium	19	2	8	8	1
Calcium	20	2	8	8	2

a hydrogen atom holds one electron, and this electron takes up residence in the first shell, *i.e.* the shell closest to the nucleus. It does not enter the second shell, because this would not result in a permanent arrangement. If the electron did enter the second shell it would immediately be pulled by the nucleus into the first shell.

Helium ($Z = 2$) has two electrons in the first shell. Lithium ($Z = 3$) has two electrons in the first shell and one in the second. All three cannot enter the first shell because, as we have seen, the first shell of an atom can never hold more than two electrons. Beryllium ($Z = 4$) has two electrons in the first shell and two in the second. These electronic arrangements of atoms are known as their *electronic configurations*. The configurations for atoms of all elements up to an atomic number of 20 are shown in Table 3.2.

Elements number 19 (potassium) and 20 (calcium) have unexpected electronic configurations, in that for these elements the fourth shell starts to fill before the third one is complete. This will be explained at a higher level of study. For the time being it is sufficient to accept that the electronic configurations of potassium and calcium, although anomalous, are *stable, i.e.* they remain as they are, and do not change to 2, 8, 9 and 2, 8, 10 respectively.

26

Summary

At the conclusion of this chapter, you should be able to:

1. recognise the order of the size and mass of an atom,
2. state that an atom consists of a positively charged nucleus with negatively charged electrons in movement around it,
3. state that protons and neutrons exist in the nucleus,
4. define 'mass number' as the total number of protons and neutrons in the nucleus,
5. state that electrons whose distances from the nucleus are approximately equal form a grouping called a shell,
6. label electron shells by the numbers 1, 2, 3, etc.,
7. state that, for any atom, the number of electrons in the shells is equal to the number of protons in the nucleus,
8. define 'atomic number' as the number of protons in the nucleus,
9. explain and illustrate the occurrence of isotopes,
10. use any suitable combination of atomic number, mass number, number of protons, neutrons and electrons to deduce the others,
11. define the 'relative mass' of a small particle,
12. state that oppositely charged particles attract each other, while similarly charged particles repel each other,
13. state that the relative charge of a proton is $+1$ and that of an electron is -1,
14. describe how the scattering of α-particles supports the theory of a nuclear atom with the protons in the nucleus,
15. write down the electronic configurations of all the elements up to an atomic number of 20.

Questions

1. Sulphur-32 is the principal isotope of sulphur (atomic number 16).
   (a) How many protons are there in the nucleus of the atom?
   (b) How many neutrons are there?
   (c) How many electrons are there outside the nucleus?
   (d) How many of these electrons are in the first shell?
   (e) How many are there in the second shell?
   (f) How many are there in the third shell?
   (g) What is the mass number of the atom?
   (h) What is the nuclear charge?
2. Which of the following statements about the relative mass of an atom are true and which are false?
   (a) It is equal to the mass number.
   (b) It is equal to the number of protons.
   (c) It is equal to the number of protons plus the number of electrons.
   (d) It is the number of times the mass of the atom is greater than that of an atom of the isotope carbon-12.
   (e) It is the mass of an atom expressed in grams.

(f) It is the number of times the mass of the atom is greater than that of an electron.

(g) It is the mass relative to that of a hydrogen atom.

(h) It is the mass relative to that of one-twelfth the mass of a carbon-12 atom.

3. Which of the following electronic configurations for atoms are permissible and which are not?

(a) 2, 8, 8.

(b) 1, 7.

(c) 2, 7, 2.

(d) 2, 9, 1.

(e) 2, 8, 8, 2.

(f) 2, 8, 10.

(g) 2, 2.

(h) 2.

In questions 4—6 select the most appropriate answer, labelled A, B, C or D.

4. If a brass sphere is negatively charged, it:

A   possesses an accumulation of electrons on its surface,

B   will attract another negatively charged sphere,

C   possesses more protons than electrons,

D   will repel a positively charged sphere.

5. Atoms of the isotopes nitrogen-14 and nitrogen-15 differ from each other in their:

A   number of protons,

B   number of neutrons,

C   number of electrons,

D   electronic configurations.

6. The isotopes of magnesium-24 and magnesium-25 have identical:

A   melting temperatures,

B   densities,

C   chemical properties,

D   mass numbers.

Chapter 4

Elements

Metals and non-metals

Elements are conveniently divided into two categories, namely *metals* and *non-metals*, partly on the basis of their physical properties and partly on chemical properties. The principal physical differences between metals and non-metals are shown in Table 4.1, while chemical differences are discussed in Chapter 10. There are a few elements, such as arsenic and germanium, which are difficult to classify in this way, because some of their features are metallic while others are non-metallic.

Ions from metals and non-metals

All atoms are electrically neutral. This is because they contain equal numbers of protons (each with a relative charge of +1) and electrons (each with a relative charge of −1). However, the electrical neutrality can easily be destroyed, by the removal or addition of electrons, to give electrically charged particles called *ions*. Atoms of metallic elements (*e.g.* sodium) are most likely to *lose* electrons to give positively charged ions, while atoms of non-metallic elements (*e.g.* chlorine) are most likely to *gain* electrons to form negatively charged ions (see Fig. 4.1). Because of this, **metals are defined as electropositive elements, and non-metals as electronegative elements.**

As a general rule, it is only the outer shell electrons which are involved when an atom is converted into an ion. In the formation of the chloride ion, for example, an electron is obliged to enter the third shell because the first two are full. In the formation of the sodium ion an electron is lost from the third shell, rather than the first or second, because the third shell is relatively far from the nucleus, and the nucleus exerts relatively little attraction at this distance from it.

Table 4.1 A comparison of the physical properties of metals and non-metals

Property	Metals	Non-metals
State	Are all solids at room temperature, except for mercury	May be gases, liquids or solids
Appearance of solid	Have a characteristic metallic lustre and (except for copper and gold) are silvery-white	Do not look like metals
Density of solid	In general, have high densities (sodium and potassium are exceptions)	In general, have low densities (bromine and iodine are exceptions)
Electrical conductivity	Are good conductors of electricity	Are poor conductors, except for carbon
Strength	Can be stretched without breaking, within certain limits	Are brittle, and break when stretched
Malleability	Are malleable, *i.e.* can be beaten into sheets	Are not malleable
Ductility	Are ductile, *i.e.* can be drawn out into wire	Are not ductile
Sonority	Are sonorous, *i.e.* give a clear, loud sound when struck	Are not sonorous

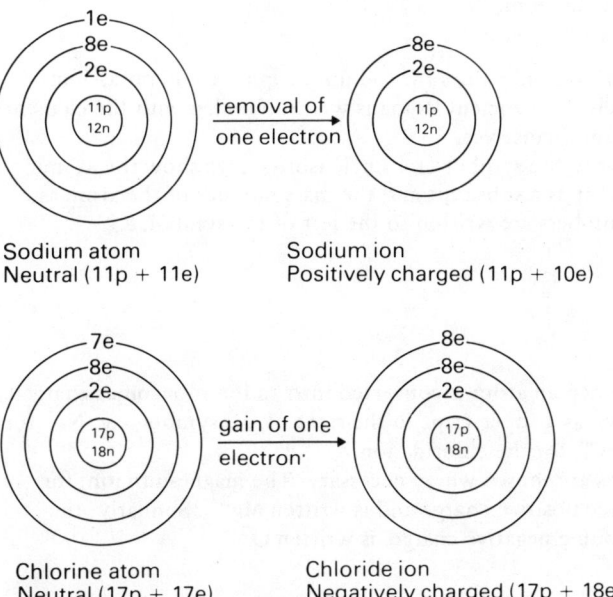

Sodium atom
Neutral (11p + 11e)

Sodium ion
Positively charged (11p + 10e)

Chlorine atom
Neutral (17p + 17e)

Chloride ion
Negatively charged (17p + 18e)

Fig. 4.1 Ion formation

More than one electron can be involved when ions are formed. An atom of magnesium, for instance, (electronic configuration 2, 8, 2) can lose two electrons to give an ion with a double positive charge, while an atom of oxygen (2, 6) can gain two electrons to yield the oxide ion with a double negative charge.

The loss or gain of electrons from an atom often takes place during chemical changes when elements react together to give compounds. For example, sodium and chloride ions are formed when sodium reacts with chlorine to give sodium chloride (see p. 48). *Ionisation*, *i.e.* the formation of ions, can also occur in *discharge tubes* when electricity, at a high voltage, is applied to an element in the gaseous state.

Symbols

Every element has a *symbol*, which represents either one atom of the element or the abbreviated name of the element. The symbols of the common elements must be memorised, and the list in Table 4.2 is provided for study. (A complete list of symbols for all the 105 known elements is provided by Fig. 4.2.) It will be seen that most symbols are abbreviations of the English names of the elements, but a few are taken from Latin names, *e.g.*

Sn, from 'stannum', for tin,
Pb, from 'plumbum', for lead,
Ag, from 'argentum', for silver.

In certain cases the symbol of an element is derived from the name of one of its compounds, *e.g.*

Na, from 'natron', for sodium,
K, from 'kalium', for potassium.

'Natron' and 'kalium' are old names for sodium carbonate and potassium carbonate respectively. The ancient Romans were unfamiliar with the elements sodium and potassium themselves.

Whenever we write the symbol of a single isotope, we show the atomic number of the element as a subscript and the mass number of the atom as a superscript. Both numbers are written to the left of the symbol, *e.g.*

mass number $\quad\quad\quad$ $^{12}_{6}\mathrm{C}$
atomic number

Ionic symbols

We have seen that when an atom is converted into an ion it becomes charged. This charge is written as a superscript to the *right* of the symbol, *e.g.* Na^+ for the sodium ion and Cl^- for the chloride ion.

Multiple charges are shown where necessary. The magnesium ion, for instance, has a double positive charge and is written Mg^{2+}. Similarly, the oxide ion, with a double negative charge, is written O^{2-}.

Table 4.2 Symbols of the common elements

Metals		Non-metals		
Element	Symbol	Element	Symbol	Appearance
Aluminium	Al	Argon	Ar	Colourless gas
Arsenic	As	Boron	B	Black solid
Barium	Ba	Bromine	Br	Red liquid
Beryllium	Be	Carbon	C	Black solid
Bismuth	Bi	Chlorine	Cl	Green gas
Cadmium	Cd	Fluorine	F	Yellow gas
Calcium	Ca	Helium	He	Colourless gas
Chromium	Cr	Hydrogen	H	Colourless gas
Cobalt	Co	Iodine	I	Shiny black solid
Copper	Cu	Krypton	Kr	Colourless gas
Gold	Au	Neon	Ne	Colourless gas
Iron	Fe	Nitrogen	N	Colourless gas
Lead	Pb	Oxygen	O	Colourless gas
Lithium	Li	Phosphorus	P	White or red solid
Magnesium	Mg	Silicon	Si	Brown solid
Manganese	Mn	Sulphur	S	Yellow solid
Mercury	Hg	Xenon	Xe	Colourless gas
Nickel	Ni			
Platinum	Pt			
Potassium	K			
Silver	Ag			
Sodium	Na			
Strontium	Sr			
Tin	Sn			
Titanium	Ti			
Vanadium	V			
Zinc	Zn			

Relative atomic mass

Almost every element occurs in nature as a mixture of isotopes, each of which has its own mass number. The average of the mass numbers in the naturally occurring mixture is known as the *relative atomic mass* (A_r) of the element. The term is defined as follows:

$$A_r = \frac{\text{average mass per atom of the naturally occurring mixture}}{\text{one-twelfth the mass of an atom of the isotope carbon-12}}$$

Relative atomic masses can be found either by experiment or by calculation. If we adopt the latter method we have to work out a weighted average of the mass numbers of the isotopes, *i.e.* an average which is weighted so as to

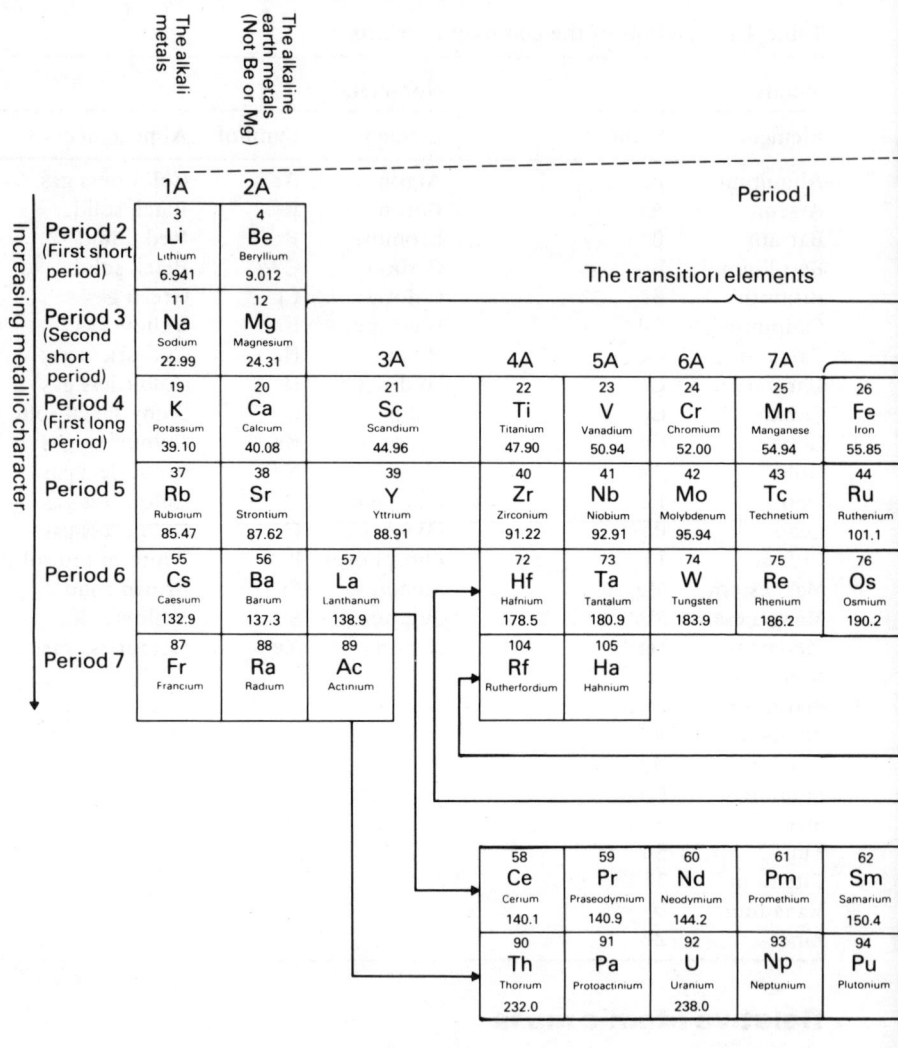

Fig. 4.2 The periodic table of the elements. The numbers above the symbols

make allowance for the proportions in which the various isotopes are present. It is not permissible merely to calculate an arithmetic mean, for this approach would give the correct answer only if all the isotopes were present in equal proportions.

For example, hydrogen consists of the isotopes hydrogen-1, hydrogen-2 and hydrogen-3. There is over 99 per cent of hydrogen-1, and the weighted average, *i.e.* the relative atomic mass of hydrogen, is 1.008. If all three isotopes were present in equal proportions, the relative atomic mass would be the arithmetic mean of 1, 2 and 3; namely 2.

| | | | | | | The chalcogens | The halogens | The noble gases |

Vertical labels: The chalcogens | The halogens | The noble gases

1								2
H Hydrogen 1.008								**He** Helium 4.003

		3B	4B	5B	6B	7B

5 **B** Boron 10.81	6 **C** Carbon 12.01	7 **N** Nitrogen 14.01	8 **O** Oxygen 16.00	9 **F** Fluorine 19.00	10 **Ne** Neon 20.18
13 **Al** Aluminium 26.98	14 **Si** Silicon 28.09	15 **P** Phosphorus 30.97	16 **S** Sulphur 32.06	17 **Cl** Chlorine 35.45	18 **Ar** Argon 39.95

8

1B	2B

27 **Co** Cobalt 58.93	28 **Ni** Nickel 58.71	29 **Cu** Copper 63.55	30 **Zn** Zinc 65.37	31 **Ga** Gallium 69.72	32 **Ge** Germanium 72.59	33 **As** Arsenic 74.92	34 **Se** Selenium 78.96	35 **Br** Bromine 79.90	36 **Kr** Krypton 83.80
45 **Rh** Rhodium 102.9	46 **Pd** Palladium 106.4	47 **Ag** Silver 107.9	48 **Cd** Cadmium 112.4	49 **In** Indium 114.8	50 **Sn** Tin 118.7	51 **Sb** Antimony 121.8	52 **Te** Tellurium 127.6	53 **I** Iodine 126.9	54 **Xe** Xenon 131.3
77 **Ir** Iridium 192.2	78 **Pt** Platinum 195.1	79 **Au** Gold 197.0	80 **Hg** Mercury 200.6	81 **Tl** Thallium 204.4	82 **Pb** Lead 207.2	83 **Bi** Bismuth 209.0	84 **Po** Polonium	85 **At** Astatine	86 **Rn** Radon

Stepwise division between metals on the left and non-metals on the right

63 **Eu** Europium 152.0	64 **Gd** Gadolinium 157.3	65 **Tb** Terbium 158.9	66 **Dy** Dysprosium 162.5	67 **Ho** Holmium 164.9	68 **Er** Erbium 167.3	69 **Tm** Thulium 168.9	70 **Yb** Ytterbium 173.0	71 **Lu** Lutetium 175.0	The lanth- anoids
95 **Am** Americium	96 **Cm** Curium	97 **Bk** Berkelium	98 **Cf** Californium	99 **Es** Einsteinium	100 **Fm** Fermium	101 **Md** Mendelevium	102 **No** Nobelium	103 **Lr** Lawrencium	The actinoids

Increasing non-metallic character →

are atomic numbers, and those below the symbols are relative atomic masses.

Chlorine consists of a mixture of approximately 75 per cent of chlorine-35 and 25 per cent of chlorine-37, which means that the relative atomic mass of chlorine is approximately 35.5. Clearly, if chlorine were a 50:50 mixture of these two isotopes, its relative atomic mass would be 36.

We can summarise this principle by a simple mathematical formula:

$$A_r = (A_1 \times f_1) + (A_2 \times f_2) + (A_3 \times f_3) + \ldots$$

where A represents the mass number of an isotope and f is the fraction in which it is present. The dots indicate that relative atomic mass is the sum of the $(A \times f)$ terms for all the isotopes in the naturally occurring mixture,

i.e. $A_r = \Sigma A \times f$, where $\Sigma = $ 'sum of'.

Most chemical calculations require a knowledge of relative atomic masses, and accurate values for all the naturally occurring elements are incorporated in Fig. 4.2 (the periodic table). The values are provided for reference purposes only and should not be memorised.

The periodic table of the elements

The *periodic table* is a way of classifying the elements so that those with similar properties are grouped together. The table is based on the observation that when the elements are placed in ascending order of atomic number, there is a periodic repetition of properties; in other words, the properties of the elements are repeated after certain intervals. To illustrate this, we shall consider the first 20 elements.

				*	**	**	***	***	****	
			*	**	**	**	***	***	****	****	*
Element	H	He	Li	Be	B	C	N	O	F	Ne
Atomic number	1	2	3	4	5	6	7	8	9	10
		*	**	**	***	***	****			*
	**	**	**	***	***	****	****	*	**	**
Element	Na	Mg	Al	Si	P	S	Cl	Ar	K	Ca
Atomic number	11	12	13	14	15	16	17	18	19	20

The first nine elements, hydrogen to fluorine, are entirely different from one another, but the tenth element, neon, closely resembles helium ($Z = 2$) in its physical and chemical properties. In the same way, sodium ($Z = 11$) resembles lithium ($Z = 3$), magnesium ($Z = 12$) resembles beryllium ($Z = 4$), and so on. Looking at the properties of the elements from neon ($Z = 10$) to chlorine ($Z = 17$) is rather like watching an old film; we have seen it all before and we know what is coming next! Other repetitions start with argon ($Z = 18$), krypton ($Z = 36$), xenon ($Z = 54$) and radon ($Z = 86$). All the elements in the list above which are related to one another have been marked with the same number of asterisks.

The related elements helium, neon, argon, krypton, xenon and radon are known collectively as the *noble gases*, formerly the 'inert gases', because they are very unreactive and are all gaseous. We can say, therefore, that there is a repetition of properties with each of the noble gases, although the repetitions begin at irregular intervals.

When compiling the periodic table we draw dividing lines after the noble gases, so that the list of elements is split up into a number of *periods* that correspond to one another to some extent. We arrange the periods one above another in such a way that elements with similar properties lie together in vertical *groups*. The principal groups are numbered 1A, 2A, 3B, 4B, 5B, 6B and 7B (see Fig. 4.2). The noble gases are not given a group number.

The elements in any group of the periodic table are related to one another for the basic reason that their atoms have the same number of electrons in their outer shells (see Table 4.3). Helium is exceptional, in that it has only two electrons in its outer shell while all the other noble gases have eight.

Table 4.3 Electronic configurations of the noble gases, the alkali metals and the halogens

The noble gases	Electronic configurations	The alkali metals (Group 1A)	Electronic configurations
Helium	2	Lithium	2,1
Neon	2,8	Sodium	2,8,1
Argon	2,8,8	Potassium	2,8,8,1
Krypton	2,8,18,8	Rubidium	2,8,18,8,1
Xenon	2,8,18,18,8	Caesium	2,8,18,18,8,1
Radon	2,8,18,32,18,8	Francium	2,8,18,32,18,8,1

The halogens (Group 7B)	Electronic configurations
Fluorine	2,7
Chlorine	2,8,7
Bromine	2,8,18,7
Iodine	2,8,18,18,7
Astatine	2,8,18,32,18,7

The reactions of an element are controlled very largely by the electrons in the outermost atomic shell. (We have already seen that, in general, it is the outer shell electrons which are involved when an atom is converted into an ion.) Elements which have the same number of electrons in their outer shells take part in similar reactions, *i.e.* they have similar chemical properties.

Molecules

It is unusual for an element to exist as independent atoms. Only the noble gases, such as argon, consist of separate atoms which move about in a random manner (see Fig. 4.3).

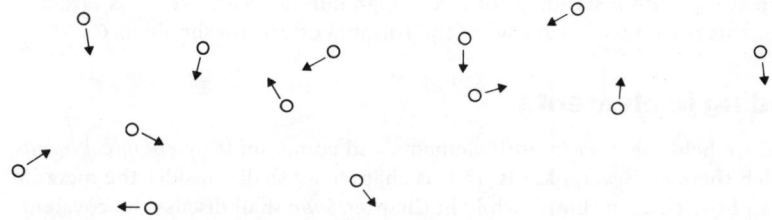

Fig. 4.3 Atoms of a noble gas

Because they are composed of single atoms, the noble gases are said to be *monoatomic*.

Atoms of the other non-metallic elements are too reactive to exist by themselves, and immediately join together in clusters of two or more to give what are known as *molecules*. An atom of nitrogen, for instance, at room temperature immediately combines with another atom of nitrogen to give a 'molecule of nitrogen', and ordinary nitrogen gas is given the *formula* N_2 to denote its molecular character.

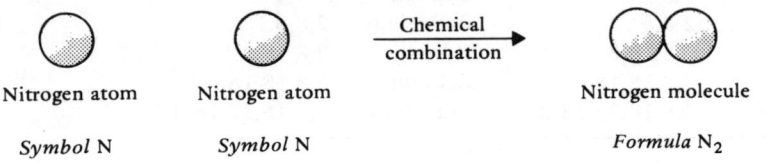

Nitrogen atom	Nitrogen atom		Nitrogen molecule
Symbol N	*Symbol* N		*Formula* N_2

Nitrogen is said to be *diatomic* because there are two atoms per molecule.

Most gaseous elements are diatomic. Hydrogen exists as H_2 molecules, chlorine as Cl_2 molecules, and oxygen is usually found as O_2 molecules. The number of atoms per molecule of an element is known as its *atomicity*.

Solid non-metals have a relatively high atomicity. Atoms of phosphorus join together to give P_4 molecules, and atoms of sulphur form S_8 molecules (see Fig. 4.4).

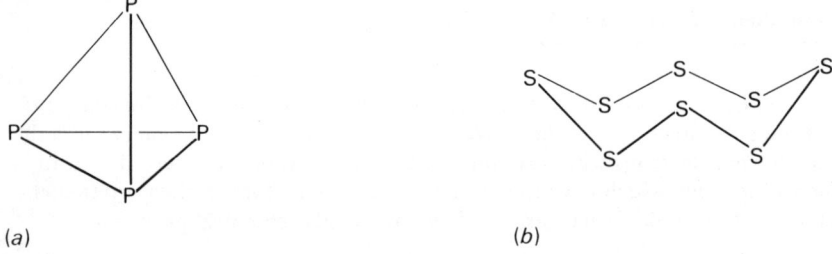

(a)   (b)

Fig. 4.4 Molecules of (*a*) white phosphorus, and (*b*) solid sulphur

Atoms of carbon, another solid non-metal, are linked together to give 'giant molecules' of indefinite size. Every piece of carbon is one enormous molecule (see Fig. 8.2), and it could be argued that the formula should be written as C_n, with n standing for a very high number. However, n is variable, and for this reason we always write the formula of carbon simply as C.

Bonding in elements

Atoms are held together in both elements and compounds by *chemical bonds*, of which there are several kinds. In this chapter we shall consider the metallic bond and the covalent bond, while in Chapter 5 we shall discuss the covalent bond in more detail and also meet the electrovalent bond. Other types of bonding will be encountered at a later stage of study.

The metallic bond

Atoms of metallic elements do not form molecules, but are packed closely together in layers which are stacked on top of one another. A metal atom is joined to other atoms in the layer, and to atoms in adjoining layers, by *metallic bonding*. Each metal atom sheds one or more electrons, in general from its outer shell, and exists as a positively charged ion. The electrons which are released in this way lie between the ions as a 'sea' (see Fig. 4.5).

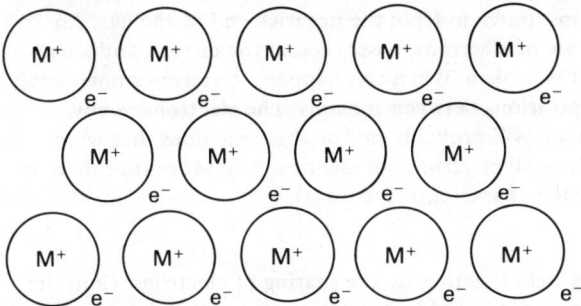

Fig. 4.5 Metallic bonding

It is wrong to suppose that an electron is associated only with the ion from which it originated. Any given electron is associated partly with the ions on its left, partly with those on its right, partly with those above, partly with those below, partly with those in front and partly with those behind. In short, an electron is 'shared' by all ions at equal distances from it.

The theory of metallic bonding accounts very well for the observed properties of a metal. Ductility and malleability are explained by the ability of layers of ions to slide over one another. Electrical conductivity is due to the ability of the electron sea to move when electricity is supplied by, say, a dry battery (see Fig. 4.6).

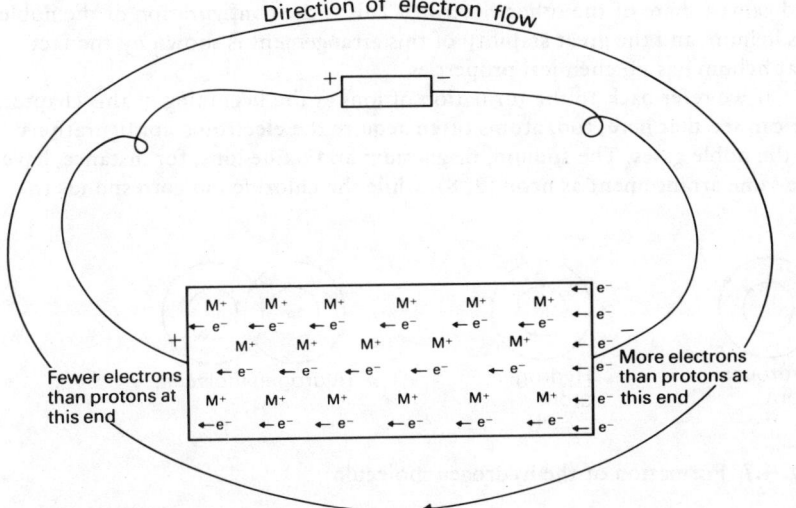

Fig. 4.6 Metallic conductivity in a metal bar

The positive pole of the battery has fewer electrons than protons (see p. 23) and shares this shortage of electrons with the end of the metal bar to which it is attached. The negative pole of the battery has a surplus of electrons and this, too, is shared with the bar at the point of contact. Consequently, when the circuit is completed, electrons immediately flow through the bar to the positive end. There is no delay while metal atoms split up into ions and electrons, because the metal is already in an ionised condition. The electrons travel out of the metal altogether, towards the battery, and their place is taken by further electrons drawn in from the negative end of the bar. An *electric current*, *i.e.* a flow of electrons, passes round the circuit, and keeps on flowing until contact is broken. When this happens, electrons immediately return to their original positions between the ions. The electrons in any particular part of the metal will probably not be the same ones that were there at the start, but since all electrons are identical to one another there is no noticeable change in the character of the metal.

The covalent bond

Atoms in a molecule are held together by the sharing of electrons. Consider, first, the combination of two atoms of hydrogen to give a molecule of hydrogen (see Fig. 4.7).

Each hydrogen atom shares its electron with the other, and the shared pair lies between the two atomic nuclei. In this way the two nuclei are bound firmly together, for each of them is attracted to the shared pair of electrons.

A shared pair of electrons which holds two atoms together is known as a *covalent bond*. Because its molecules possess covalent bonds, hydrogen gas is a *covalent substance*. All substances which consist of molecules are said to be 'covalent' for the same reason.

Atoms form covalent bonds because, in doing so, they acquire a stable electronic configuration. In the hydrogen molecule, each atom in effect acquires two electrons in its first shell. (It retains a share of its own electron, and gains a share of the other electron.) This is the configuration of the noble gas helium, and the great stability of this arrangement is shown by the fact that helium has no chemical properties.

If we refer back to the formation of ions at the beginning of this chapter, we can see that here, too, atoms often acquire the electronic configurations of the noble gases. The sodium, magnesium and oxide ions, for instance, have the same arrangement as neon (2, 8), while the chloride ion corresponds to

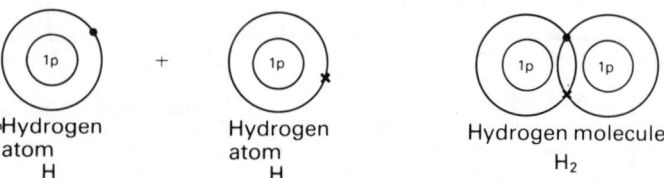

Hydrogen atom
H

Hydrogen atom
H

Hydrogen molecule
H_2

Fig. 4.7 Formation of the hydrogen molecule

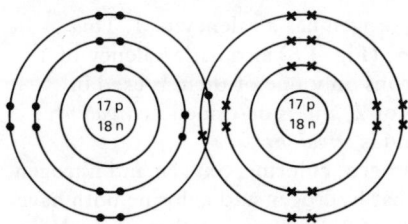

Fig. 4.8 The chlorine molecule

argon (2, 8, 8). It is a general principle that atoms often (but not always) achieve noble gas configurations when they react together. We shall discuss this in more detail in the next chapter.

The bonding in a chlorine molecule closely resembles that in a hydrogen molecule. The two chlorine atoms are held together by a single covalent bond, formed by each atom sharing one of its outer shell electrons with the other (see Fig. 4.8). Each atom of chlorine in effect acquires the configuration 2, 8, 8; *cf.* the noble gas argon.

The oxygen molecule is rather more complicated, in that the atoms are joined together by a *double bond, i.e.* two covalent bonds lying side by side. They are formed by each oxygen atom sharing *two* of its outer shell electrons with the other (see Fig. 4.9). Each oxygen atom acquires the configuration 2, 8; *cf.* neon.

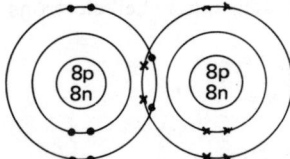

Fig. 4.9 The oxygen molecule

In the nitrogen molecule, N_2, there is a *triple bond, i.e.* three covalent bonds, between the atoms (see Fig. 4.10). Again, the neon configuration (2, 8) is achieved.

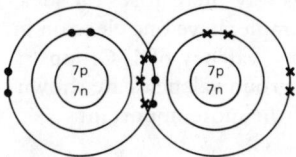

Fig. 4.10 The nitrogen molecule

Valency

Apart from hydrogen, elements never use all their atomic electrons in forming chemical bonds. The number of electrons that actually takes part in bonding is a characteristic of an element and is known as its *valency*.

If we look at Fig. 4.7, we see that hydrogen has a valency of 1. This element is said to be *monovalent*. Chlorine (Fig. 4.8) also has a valency of 1, for although the atom possesses 17 electrons only one of them is used in bonding. Oxygen (Fig. 4.9) has a valency of 2, and is described as *divalent*. Nitrogen (Fig. 4.10) has a valency of 3 and is *trivalent*.

Because, in the above examples, hydrogen, chlorine, oxygen and nitrogen are forming covalent bonds, we can say that hydrogen and chlorine both have a 'covalency of 1', oxygen has a 'covalency of 2', and nitrogen has a 'covalency of 3'. The *covalency* of any element can be defined as the number of covalent bonds that are formed by an atom of that element.

Variable valency

Some elements have a constant valency, but others have different valencies in different substances. For example, hydrogen always has a valency of 1, but iron has a valency of 2 in some of its compounds and 3 in others. Such an element is said to exhibit *variable valency*.

Maximum covalency

Except for the transition elements (see Fig. 4.2), only the electrons of the outer shell of an atom are utilised in bonding. We cannot say that all the outer electrons will necessarily be used in bonding, but they *may* be drawn upon for this purpose. It follows that, in theory, the maximum covalency of an element is equal to the number of electrons in the outer shell of its atoms. This principle is only a guide, and it must be stressed that there are some elements, notably the following, which never use all their outer shell electrons in bonding:

 i. nitrogen, which always has a valency of 3,
 ii. oxygen, which always has a valency of 2,
iii. fluorine, which always has a valency of 1.

Valency and the periodic table

It is essential for any serious student of chemistry to learn the valencies of the common elements, a task which is best approached via the periodic table. Valency, like the other properties of an element, is dependent on atomic structure, and since the various members of a group have related atomic structures they usually have similar valencies. In this way there arises the idea of *group valency*. All elements of group 1A, for example, have one electron in their outermost shell and consequently they all have a valency of 1. Group 1A is said to have a 'group valency of 1'. The other group valencies are shown in Fig. 4.11, which also gives the actual valencies of the more important elements.

The valencies of the principal transition elements, and also of zinc, are as follows:

Element	Fe	Cu	Ag	Zn
Valencies	2, 3	1, 2	1	2

Group	1A	2A	3B	4B	5B	6B	7B
Group valencies	1 only	2 only	3 only	2 and 4	3 and 5	2, 4 and 6	usually 1

	Li 1	(Be)	B 3	C 4	N 3	O 2	F 1
Elements and actual valencies	Na 1	Mg 2	Al 3	Si 4	P 3,5	S 2,4,6	Cl 1,3,5,7
	K 1	Ca 2	(Ga)	(Ge)	(As)	(Se)	Br 1,3,5,7
	Rb 1	Sr 2	(In)	Sn 2,4	(Sb)	(Te)	I 1,3,5,7
	Cs 1	Ba 2	(Tl)	Pb 2,4	(Bi)	(Po)	(At)
	(Fr)	(Ra)					

H 1

Notes
1. The less common elements are enclosed by brackets.
2. Valencies of minor importance are shown in light type.

Fig. 4.11 Valencies of the common non-transition elements

Summary

At the conclusion of this chapter, you should be able to:
1. identify the important differences between metals and non-metals in terms of appearance, electrical conductivity, density, melting temperature and boiling temperature,
2. explain the electrical neutrality of the atom,
3. state that an atom can be converted into an ion by the loss or gain of one or more electrons,
4. state that, in general, it is only the outer shell electrons which are involved in ion formation,
5. state that metals form positively charged ions and non-metals form negatively charged ions,
6. state the symbols of the common elements, given their names,
7. define the 'relative atomic mass' of an element,
8. state that in the periodic table the elements are arranged in increasing order of atomic number,
9. state that elements in any group of the periodic table have the same number of electrons in their outer shells and hence similar properties,
10. state that atoms of non-metallic elements can join together to give molecules,
11. describe the metallic bond in terms of positive ions in a 'sea' of electrons,
12. describe the covalent bond in a molecule in terms of the mutual attraction of atomic nuclei for shared electrons,

13. explain the electrical conductivity of a metal in terms of the presence of mobile electrons,
14. define the 'valency' of an element as the number of electrons which are involved in bonding,
15. state the valencies of the common elements.

Questions

1. Which of the following elements are metals and which are non-metals?
   - (a) Sulphur
   - (b) Sodium
   - (c) Iodine
   - (d) Calcium
   - (e) Silicon
   - (f) Bismuth
   - (g) Phosphorus
   - (h) Argon

2. Write down the symbols of the following elements:
   - (a) Argon
   - (b) Aluminium
   - (c) Sodium
   - (d) Strontium
   - (e) Tin
   - (f) Titanium
   - (g) Magnesium
   - (h) Manganese

3. Write down the names of the elements whose symbols are as follows:
   - (a) K
   - (b) Pb
   - (c) Ag
   - (d) As
   - (e) Au
   - (f) Fe
   - (g) Cr
   - (h) Hg

4. Write down the formulae of the following molecules:
   - (a) Chlorine
   - (b) Oxygen
   - (c) Nitrogen
   - (d) Phosphorus
   - (e) Carbon

5. Which of the following elements exhibit variable valency?
   - (a) Sodium
   - (b) Tin
   - (c) Iron
   - (d) Aluminium
   - (e) Calcium
   - (f) Sulphur
   - (g) Oxygen
   - (h) Copper

   In questions 6–13 select the most appropriate answer, labelled A, B, C or D.

6. A sulphide ion is derived from an atom of sulphur (atomic number 16) by the gain of two electrons. A sulphide ion therefore possesses:
   - A   a double positive charge,
   - B   the electronic configuration 2, 8, 8,
   - C   an atomic number of 14,
   - D   an atomic number of 18.

7. Relative atomic mass has:
   - A   units of grams,
   - B   units of kilograms,
   - C   any SI recognised units,
   - D   no units.

8. Copper consists of approximately 70 per cent of the isotope copper-63

and 30 per cent of the isotope copper-65. Its relative atomic mass is therefore approximately:

A   63.0,
B   63.5,
C   64.0,
D   64.5.

9. The periodic table is compiled by arranging elements in increasing order of:

A   atomic number,
B   relative atomic mass,
C   number of neutrons in the nucleus,
D   mass number of the most abundant isotope.

10. Elements in the same group of the periodic table have similar properties because:

A   they are all either metals or non-metals,
B   their atoms contain the same number of electrons,
C   the same number of electron shells are occupied,
D   they have the same number of electrons in the outer shell of their atoms.

11. When a dry battery is connected to a bar of metal electricity flows through the circuit. Which one of the following statements is *untrue*?

A   Electrons flow from the negative pole of the battery to the positive pole via the rest of the circuit.
B   The passage of electricity causes no permanent change in the nature of the metal.
C   When electricity flows the metal ions are stationary; only the electrons move.
D   On completion of the circuit, the dry battery causes the metal atoms to form ions and electrons.

12. Which one of the following statements about covalent bonding between two atoms X and Y is *untrue*?

A   The shared pair of electrons is concentrated in the region between the two atomic nuclei.
B   The shared pair of electrons binds the atomic nuclei together by attracting both of them.
C   Both the electrons of the shared pair originate from one atom.
D   More than one covalent bond can exist between X and Y.

13. The valency of an element is defined as the:

A   total number of electrons in the atom,
B   number of electrons in the outer shell of the atom,
C   number of electrons in the first shell of the atom,
D   number of electrons that takes part in bonding.

Chapter 5

Compounds

Compounds are formed by the chemical combination of different elements.
We have already seen that hydrogen and oxygen combine to give water, that
sodium and chlorine give sodium chloride, and that mercury and oxygen give
mercury oxide. More than two elements may be involved; for example,
sulphuric acid is a chemical compound of hydrogen, sulphur and oxygen.

In most compounds the atoms of the contributing elements are present
in a simple, whole number ratio. In water, for example, the ratio of hydrogen
atoms to oxygen atoms is 2:1, and in sodium chloride the ratio of sodium
atoms to chlorine atoms is 1:1. Furthermore, for any compound this ratio is
fixed, regardless of the method by which the compound is prepared. Water,
for instance, can be made by the direct combination of hydrogen and oxygen,
by burning petrol in air, by heating sodium hydrogencarbonate ('baking
powder') or by a host of other methods, but in all cases the product is the
same. In other words, water, like all compounds, has a constant composition.

It can happen that the atoms of two or more given elements join together
in more than one ratio. In such cases a variety of compounds is formed, one
for each combining ratio. For example, the combination of hydrogen and
oxygen atoms in the ratio 2:1 gives water, but combination in the ratio 1:1
gives an entirely different compound, called hydrogen peroxide.

Two main types of compound are recognised, namely:
 i. *molecular compounds*, which are composed of molecules, and
 ii. *ionic compounds*, which are composed of ions.
We shall consider each in turn.

Molecular compounds (covalent compounds)

We have seen in Chapter 4 that two or more atoms of the same non-metallic element can join together by covalent bonds, *i.e.* shared pairs of electrons, to give a molecule of that element. In the same way, covalent bonds can form between two or more atoms of different non-metallic elements to give a molecule of a compound. For example, the formation of a covalent bond between an atom of hydrogen and an atom of chlorine gives a molecule of hydrogen chloride (see Fig. 5.1). The hydrogen atom shares its electron with the chlorine atom, and the chlorine atom in turn shares one of its outer shell electrons with the hydrogen atom. As in the case of the H_2 or the Cl_2 molecule, the shared pair of electrons is located principally between the two atomic nuclei, which are thereby bound together.

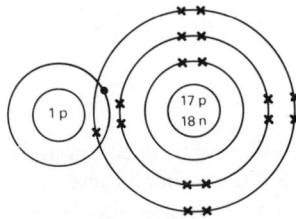

i.e.   H—Cl

Fig. 5.1 The hydrogen chloride molecule

Hydrogen and chlorine combine in an atomic ratio of 1:1 because each of these elements has a valency (see p. 39) of 1. A hydrogen atom possesses only one electron, and a chlorine atom, although it has seven electrons in its outer shell, normally uses only one for bonding purposes. The remaining six exist in pairs, known as *lone pairs* of electrons.

Oxygen, however, has a valency of 2, *i.e.* two electrons in the outer shell of the atom are available for bonding. This valency must be satisfied. Thus, when hydrogen and oxygen atoms join together it is impossible to obtain a molecule H—O. (The line denotes a covalent bond.) Instead, we get molecules of the type H—O—H (water) or H—O—O—H (hydrogen peroxide), in which every oxygen atom forms two covalent bonds. It will be seen that the covalency of hydrogen is always 1.

These examples show that an atom of oxygen can either use both of its valency electrons in bonding to atoms of a single element (see Fig. 5.2(a)), or can use one electron in bonding to an atom of one element and the second in bonding to an atom of another (see Fig. 5.2(b)). Oxygen is not alone in this respect. Atoms of any element with a valency greater than 1 can use their electrons in forming bonds to atoms of more than one element. Carbon, for instance, has a valency of 4, and can form covalent bonds with as many as

46

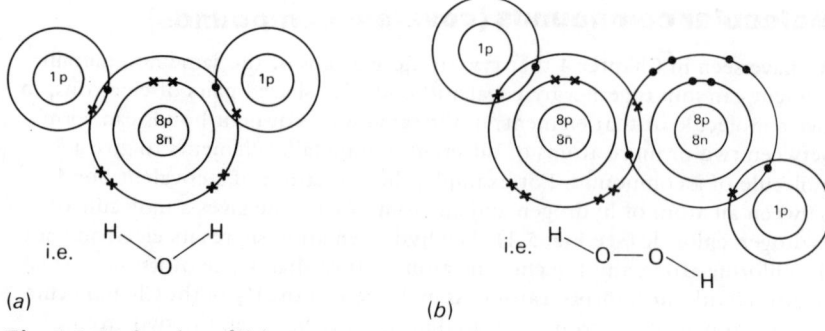

Fig. 5.2 Molecules of (*a*) water, and (*b*) hydrogen peroxide

four different elements, *e.g.*

$$\begin{array}{cccc}
\text{H} & \text{H} & \text{H} & \text{H} \\
| & | & | & | \\
\text{H}-\text{C}-\text{H} & \text{H}-\text{C}-\text{Cl} & \text{H}-\text{C}-\text{Cl} & \text{I}-\text{C}-\text{Cl} \\
| & | & | & | \\
\text{H} & \text{H} & \text{Br} & \text{Br}
\end{array}$$

Multiple covalent bonds (*i.e.* double or triple bonds) occur in compounds just as they do in elements, as the examples in Fig. 5.3 show. In all these compounds the valencies given in Fig. 4.11 are obeyed. Hydrogen displays a valency of 1, oxygen 2, carbon 4, and sulphur its maximum value of 6.

sulphuric acid   ethanoic acid   ethyne

Fig. 5.3 Structural formulae of some molecular compounds

It is important that you become adept at drawing 'dot and cross' diagrams (like those shown in Figs. 5.1 and 5.2) for common molecules. For practice, you should draw such diagrams to show the bonding in molecules of ammonia, $\begin{array}{c}\text{H} \\ | \\ \text{N} \\ / \quad \backslash \\ \text{H} \quad \text{H}\end{array}$, and methane, $\begin{array}{c}\text{H} \\ | \\ \text{H}-\text{C}-\text{H} \\ | \\ \text{H}\end{array}$. (The atomic numbers of nitrogen and carbon are 7 and 6 respectively.)

Simple molecular compounds can be recognised by the fact that they are invariably gases, liquids or solids with a low melting temperature (below approximately 300 °C (573 K)). Hydrogen chloride, for example, is a gas, water is a liquid and glucose is a solid with a melting temperature of 146 °C (419 K). These and other properties are discussed in Chapter 8.

Giant molecules

We saw in Chapter 4 that the element carbon exists not as small, discrete molecules, but as 'giant molecules' of indefinite size. Certain compounds, in a similar manner, form giant molecules in which atoms of two or more elements are held together by covalent bonds as chains or three dimensional networks which extend throughout the entire mass of the substance. Silicon disulphide, for instance, has a chain structure:

Silica (sand, quartz) has a network structure (see Fig. 8.2).

Compounds of this sort are very different from those which are composed of simple molecules. They are never gases, liquids or low melting temperature solids, but are all solid substances with extremely high melting temperatures. Silica, for example, has a melting temperature of approximately $1700\,^{\circ}C$ (1973 K). The reason is given in Chapter 8.

Formulae of molecular compounds

For every simple molecular compound we can write a *molecular formula*, which is a set of atomic symbols and subscript numbers designed to show, at a glance, the composition of the molecule. The formula tells us not merely what elements are present, but also how many atoms of each element there are in one molecule of the compound. For example, water has the formula H_2O, to indicate that in a molecule of water there are two atoms of hydrogen and one of oxygen.

The molecular formulae of the other compounds mentioned above are as follows:

HCl	H_2O_2	H_2SO_4
hydrogen chloride	hydrogen peroxide	sulphuric acid
$C_2H_4O_2$	$C_6H_{12}O_6$	C_2H_2
ethanoic acid	glucose	ethyne

For compounds whose molecules contain only a few atoms, such as water and sulphuric acid, molecular formulae are in common use, but in cases where the molecular structure is more complicated, *e.g.* ethanoic acid, *structural formulae* of the type shown in Fig. 5.3 are usually preferred. Not only do they provide more information on the way in which the atoms are joined together, but they also avoid the ambiguity that can arise whenever two or more compounds share the same molecular formula. This phenomenon, known as *isomerism*, is illustrated by butane and methylpropane, both of which have the molecular formula C_4H_{10} :

butane   methylpropane

Molecular formulae are not written for compounds which form giant molecules, because the numbers involved would be extremely high and would vary from one sample of a substance to another. (Remember that every piece of the substance is one enormous molecule.) For such compounds we write an *empirical formula*, to show the ratio in which the atoms are present. For example, the empirical formula of silicon disulphide is SiS_2, because the ratio of silicon atoms to sulphur atoms in the chains shown earlier is $1:2$. Silica (see Fig. 8.2) has an empirical formula of SiO_2.

Unfortunately, empirical formulae like these are easily mistaken for molecular formulae, and can give the false impression that silicon disulphide and silica are composed of SiS_2 and SiO_2 molecules respectively. To avoid this confusion, we can use the letter n to denote an unspecified high number, and write 'molecular formulae' of the style $(SiS_2)_n$ and $(SiO_2)_n$.

Ionic compounds (electrovalent compounds)

So far, in this chapter, we have restricted the discussion to covalent bonds formed between two atoms of non-metallic elements. In such cases the shared pair of electrons is located in the region between the two atomic nuclei, because both atoms exert a roughly equal attraction for electrons. But let us now consider what would happen if a covalent bond was formed between an atom of a non-metal and one of a metal. We have already seen that non-metals are electronegative, while metals are electropositive (see p. 28). Consequently, in this case, the shared pair of electrons does not remain between the two nuclei, but is pulled towards the non-metal atom. In effect, the bonding electron of the non-metal atom stays put, while that of the metal atom is transferred to the non-metal atom. Because an electron is negatively charged, its transfer in this way causes the non-metal atom to be converted into a negative ion, while the metal atom remains as a positive ion.

Consider, for example, the electron transfer that occurs when an atom of sodium reacts with one of chlorine in the formation of sodium chloride.

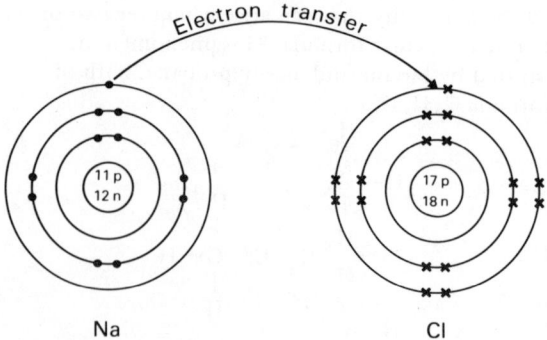

Fig. 5.4 The formation of sodium chloride

The sodium atom acquires a single positive charge and becomes a sodium ion, Na^+, while the chlorine atom gains a single negative charge and becomes a chloride ion, Cl^-. The two atomic nuclei are completely unaffected by the conversion from atoms to ions. The underlying electron shells, *i.e.* those between the nucleus and the outer shell, are also unaltered in content, but suffer a marked change in size.

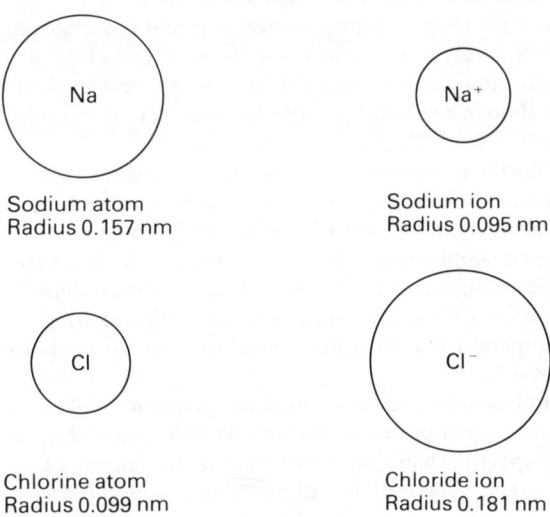

Sodium atom
Radius 0.157 nm

Sodium ion
Radius 0.095 nm

Chlorine atom
Radius 0.099 nm

Chloride ion
Radius 0.181 nm

Fig. 5.5 Change of size on ion formation

A sodium ion (radius 0.095 nm) is much smaller than a sodium atom (radius 0.157 nm), partly because the third shell is vacant, and partly because, in the ion, the ten electrons in the shells are pulled closely together by the action of eleven protons in the nucleus. By contrast, a chloride ion (0.181 nm) is much larger than a chlorine atom (0.099 nm), despite the fact that the same number of electron shells is occupied in both. The reason is that, on ionisation, the number of electrons is increased from 17 to 18. Electrons in the ion therefore repel one another more than in the atom, and move further from the nucleus to reduce the repulsion.

The bonding in sodium chloride is described as *electrovalent*. The term 'electrovalent bonding' means the *transfer* of one or more electrons from one atom to another, in contrast to 'covalent bonding', which, as we have seen, means the *sharing* of a pair of electrons between two atoms.

Electrovalent bonding is merely an extreme case of covalent bonding, and there is no sharp line of demarcation between the two. A bond pair of electrons can occupy an infinite variety of positions between two atomic nuclei A and B, so that an infinite variety of bonding is possible between two extremes represented by $A^- B^+$ and $A^+ B^-$.

$$A\overset{\cdot}{\underset{\cdot}{x}}\ B \qquad A\ \overset{\cdot}{x}\ B \qquad A\ \overset{\cdot}{x}\ B \qquad A\ \overset{\cdot}{x}\ B \qquad A\ \overset{\cdot}{x}B$$

i.e. $\ A^-\ \ \ B^+ \qquad\qquad\qquad\qquad\qquad\qquad\qquad A^+\ \ B^-$.

   (i)   (ii)   (iii)   (iv)   (v)

In any particular compound, only *one* of these arrangements is adopted.

    The bonding in situations (i) and (v) is described as 'pure electrovalency', that in situation (iii) is 'pure covalency', and that in situations (ii) and (iv) is 'polar covalency'. Polar covalency is treated at a higher level of study.

    Despite this gradation of bonding, it is both convenient and theoretically sound to classify compounds simply as covalent (or molecular) and electrovalent (or ionic). Sodium chloride, for instance, is said to be an electrovalent (or ionic) compound, even though the bond pair of electrons does not reside entirely at the chlorine atom.

    A crystal of sodium chloride consists of a symmetrical arrangement of sodium ions and chloride ions (see Fig. 8.3). The ratio of ions is 1:1, which produces an electrically neutral crystal. **It must be very clearly understood that there is no such thing as a simple molecule of sodium chloride.** It is very difficult to disrupt the ionic arrangement by means of heat, so that sodium chloride, in common with other ionic compounds, has a very high melting temperature and boiling temperature. (The properties of ionic compounds are fully discussed in Chapter 8.)

    An atom of sodium can lose only one electron in forming ions, and an atom of chlorine can gain only one. Both elements are said to have an 'electrovalency of 1'. (This is more specific than simply saying that the valency of each of these elements is 1.) The *electrovalency* of an element is defined as the number of electrons lost or gained by an atom of that element in the formation of ions.

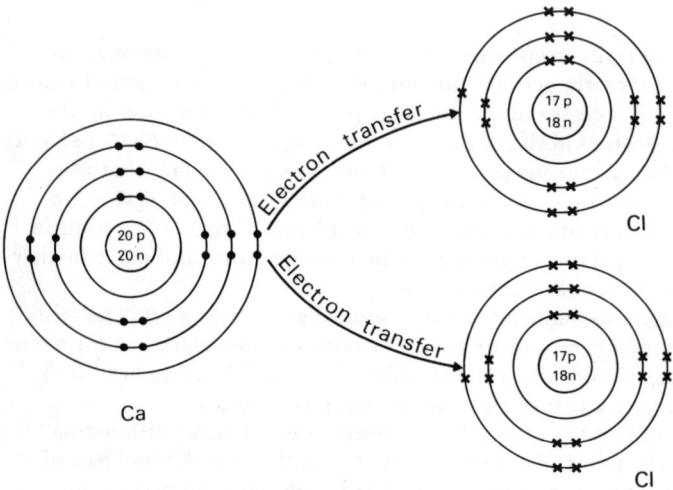

Fig. 5.6 The formation of calcium chloride

It is worth noting that when chlorine combines with other non-metals (*e.g.* hydrogen) it has a covalency of 1, but when it combines with metals it has an electrovalency of 1. Sodium, by contrast, always exhibits an electro-valency of 1. Some metals, *e.g.* aluminium, tin and lead, are able to form covalent bonds as well as electrovalent bonds, but metals in groups 1A and 2A of the periodic table display electrovalent bonding to the almost complete exclusion of covalent bonding.

Calcium has an electrovalency of 2, *i.e.* the atom, in forming a calcium ion, loses two outer shell electrons. Thus, when calcium reacts with chlorine to form calcium chloride, each calcium atom requires *two* atoms of chlorine to receive these electrons (see Fig. 5.6).

A crystal of calcium chloride is thus composed of a symmetrical arrangement of Ca^{2+} and Cl^- ions in the ratio 1:2. Again we see that the crystal as a whole is electrically neutral.

Noble gas configurations

There are six 'noble gases', and as we have seen (Fig. 4.2) they constitute a vertical group of the periodic table. The word *noble* in chemistry means 'inert', *i.e.* unreactive, and the noble gases are so called because they participate in very few reactions. In fact, the first three, namely helium, neon and argon, have virtually no chemistry at all. This is because atoms of noble gases are stable, in the sense that it is difficult to remove electrons from them and just as difficult to introduce any more.

When an atom reacts with other atoms, it often acquires the electronic configuration of one of these noble gases (see Table 5.1). The reason is that whenever a change occurs in nature, the materials undergoing change strive to achieve stability, *i.e.* the most stable possible state. (That is why water, when allowed to do so, flows downhill.) Compounds in which all the atoms or ions have acquired noble gas configurations are stable, because it is not easy for the electrons to take part in further bond formation.

Many simple ions, *e.g.* Na^+, Ca^{2+} and Cl^-, are *isoelectronic* with the noble gases, *i.e.* they have the same electronic configurations as atoms of the noble gases. Examples are shown in Table 5.2.

Most common ions, except for those derived from tin, lead, antimony, bismuth and many of the transition elements, possess noble gas configurations.

Table 5.1 Electronic configurations of the noble gases

Noble gas	Symbol	Electronic configuration
Helium	He	2
Neon	Ne	2, 8
Argon	Ar	2, 8, 8
Krypton	Kr	2, 8, 18, 8
Xenon	Xe	2, 8, 18, 18, 8
Radon	Rn	2, 8, 18, 32, 18, 8

Table 5.2 Electronic configurations of certain ions

Element	Electronic configuration		Noble gas for comparison
	in the atom	in the ion	
Sodium	2,8,1	2,8	Neon
Calcium	2,8,8,2	2,8,8	Argon
Chlorine	2,8,7	2,8,8	Argon
Hydrogen	1	2	Helium

Table 5.3 Examples of ion formation

Element	Electronic configuration	Nearest noble gas	Number of electrons that the atom must lose or gain to reach the nearest noble gas	Resultant ion
Potassium	2,8,8,1	Argon	Lose 1	K^+
Aluminium	2,8,3	Neon	Lose 3	Al^{3+}
Lithium	2,1	Helium	Lose 1	Li^+
Magnesium	2,8,2	Neon	Lose 2	Mg^{2+}
Bromine	2,8,18,7	Krypton	Gain 1	Br^-
Oxygen	2,6	Neon	Gain 2	O^{2-}
Sulphur	2,8,6	Argon	Gain 2	S^{2-}
Nitrogen	2,5	Neon	Gain 3	N^{3-}

With this knowledge it is easy for us to predict the nature of the ion that is likely to be formed by an element when it enters into electrovalent bonding (see Table 5.3).

Noble gas configurations are not always achieved when atoms react together to give covalent compounds, although this does apply for many simple molecules, such as H_2O, Cl_2, HCl, CH_4 and NH_3.

Covalent ions

Many common ions are derived not from a single atom, but from two or more atoms bonded together covalently. A simple example is the ammonium ion, $NH_4{}^+$:

$$\left[\begin{array}{c} H \\ | \\ \diagdown N \diagup \\ H \; | \; H \\ H \end{array} \right]^+$$

As usual, the short lines represent covalent bonds. The ion as a whole has a single positive charge because it contains one more proton than electron.

Covalent ions may be either positively or negatively charged, but $NH_4{}^+$ is

Table 5.4 Common negatively charged covalent ions

Uninegative ions		Binegative ions	
Formula	Name	Formula	Name
CN^-	Cyanide	$CO_3{}^{2-}$	Carbonate
CH_3COO^-	Ethanoate or acetate	$SO_4{}^{2-}$	Sulphate
$HCO_3{}^-$	Hydrogencarbonate (formerly 'bicarbonate')	$SO_3{}^{2-}$	Sulphite
$HSO_4{}^-$	Hydrogensulphate (formerly 'bisulphate')	**Trinegative ions**	
OH^- †	Hydroxide	Formula	Name
ClO^-	Hypochlorite		
$NO_3{}^-$	Nitrate	$PO_4{}^{3-}$	Phosphate
$NO_2{}^-$	Nitrite		

† IUPAC recommends writing the formula of this ion as HO^-.

the only positive one which is likely to be met at this level. Some of the more common negatively charged covalent ions are shown in Table 5.4.

All the ions shown in Table 5.4 (except CN^-) contain combined oxygen and are known as *oxoions*. In cases where more than one oxoion is derived from a particular element, *e.g.* nitrogen or sulphur, the ion of relatively low oxygen content has a name which ends in '-ite', while the ion of relatively high oxygen content has a name ending in '-ate'. Compare, for example, nitrite with nitrate, and sulphite with sulphate. See also the naming of acids, page 56.

Formulae of ionic compounds

Ionic compounds cannot be represented by molecular formulae for the simple reason that they do not consist of molecules. Instead, we write empirical formulae to show the ratio in which the ions are present. Thus, the formula of sodium chloride is NaCl. If we wish to emphasise that sodium chloride is ionic and not molecular we can write the formula as $Na^+ Cl^-$. For calcium chloride, where the ratio of calcium ions to chloride ions is 1:2, the formula is $CaCl_2$, *i.e.* $Ca^{2+} (Cl^-)_2$.

The presence of covalent ions does not affect the manner in which the formula is written. For example, the compounds potassium nitrate and ammonium chloride, in both of which the ionic ratio is 1:1, have the formulae KNO_3 and NH_4Cl respectively. Sodium carbonate, in which the ratio of sodium ions to carbonate ions is 2:1, has the formula Na_2CO_3.

If the proportion of covalent ions is greater than that of simple ions, the covalent ion is enclosed in brackets and the appropriate multiplying factor is written outside the right-hand bracket as a subscript. For example, calcium hydroxide, which contains calcium ions and hydroxide ions in the ratio 1:2, has the formula $Ca(OH)_2$. As in mathematics, everything inside the brackets

Table 5.5 The formulae of some ionic compounds

Compound	Ions present	Ratio of positive ions to negative ions to secure neutrality	Formula
Potassium iodide	K^+ I^-	1 : 1	KI
Sodium sulphate	Na^+ $SO_4{}^{2-}$	2 : 1	Na_2SO_4
Ammonium nitrate	$NH_4{}^+$ $NO_3{}^-$	1 : 1	NH_4NO_3
Aluminium fluoride	Al^{3+} F^-	1 : 3	AlF_3
Potassium oxide	K^+ O^{2-}	2 : 1	K_2O
Magnesium hydroxide	Mg^{2+} OH^-	1 : 2	$Mg(OH)_2$
Aluminium sulphate	Al^{3+} $SO_4{}^{2-}$	2 : 3	$Al_2(SO_4)_3$

is multiplied by the number outside; thus, the ratio of calcium to oxygen to hydrogen atoms in calcium hydroxide is 1:2:2. In ammonium sulphate the ratio of ammonium ions to sulphate ions is 2:1, and the formula is $(NH_4)_2SO_4$.

We have previously remarked that crystals of all ionic compounds are electrically neutral. Consequently, when writing their formulae, we must ensure that the number of positive charges is equal to the number of negative charges. The examples in Table 5.5 show how this rule is applied.

In more difficult cases, *e.g.* calcium phosphate, it may be helpful to adopt the following approach. First, write down the formulae of the two ions:

$$Ca^{2+} \qquad PO_4{}^{3-}$$

Then multiply the positive ion (Ca^{2+}) by the *charge number* of the negative ion, *i.e.* the value of the negative charge. This gives us $(Ca^{2+})_3$. Similarly, multiply the negative ion by the charge number of the positive ion: $(PO_4{}^{3-})_2$. The formula of calcium phosphate is thus $(Ca^{2+})_3 (PO_4{}^{3-})_2$, *i.e.* $Ca_3(PO_4)_2$, which is electrically neutral.

Classification and nomenclature of compounds

Whenever in science we have to study a large number of items we *classify* them, *i.e.* we sort them out into 'classes' or groups with similar features, and we study the characteristics of each class. Often we cannot study all the individual members of a class, simply because life is too short! This approach is widely adopted in chemistry, and all substances are classified into a handful of different types. In this chapter we shall mention the more important classes, together with the rules for their *nomenclature* (*i.e.* naming) as laid down by the International Union of Pure and Applied Chemistry (IUPAC). Later in the book we shall discuss the chemistry of some of these classes.

It is most important for all students of chemistry to develop the habit of classifying substances, for this is how a professional chemist works. Whenever he meets a new substance, a chemist asks himself, 'What type of substance is it?' In other words, he classifies it; and if he does so correctly he can understand and even predict a great deal about the chemical properties of that substance.

Compounds are classified primarily into two broad groups, namely *organic* and *inorganic*. Organic compounds, generally speaking, are those which contain carbon. Examples are methane, CH_4, ethanol, C_2H_5OH, and ethanoic acid, CH_3COOH. Such substances are termed 'organic' because many of them are obtainable from plants or animals. Inorganic compounds comprise all non-carbon containing substances, such as water, sodium chloride and silica. The distinction between organic and inorganic compounds has no theoretical basis, nor is it particularly rigid, for a few carbon containing substances (notably carbonates) are studied in inorganic chemistry. The division is merely one of convenience, and stems from the existence of a tremendous number of carbon containing compounds.

Inorganic and organic compounds are both subdivided into various classes. We shall consider each group in turn.

Inorganic compounds

Binary compounds

Binary compounds are those which are derived from two elements only. Their names are constructed by writing down the names of the two constituent elements, in the order in which they appear in the formula, but altering the second name so that it ends in '-ide'. Thus, if oxygen appears last in the formula, the compound is called an oxide. If chlorine appears last the compound is a chloride, if hydrogen appears last the compound is a hydride, and so on.

Examples of binary compounds which are derived from a metal and a non-metal are as follows:

CaO calcium oxide $AlCl_3$ aluminium chloride

In both these examples the metal has a constant valency, which need not be specified in the name. However, if the metal has a variable valency it will probably form more than one binary compound with a given non-metal. In such cases the valency of the metal is shown by a Roman numeral enclosed in brackets immediately after its name. Tin, for example, has two valencies, namely 2 and 4, and forms two chlorides, $SnCl_2$ and $SnCl_4$. They are called tin(II) chloride and tin(IV) chloride respectively.

Binary compounds formed between two non-metals are usually named on a different system, by which the proportions of the two non-metals are indicated by prefixes of Greek origin, *e.g.* mono, di, tri, tetra, penta and hexa. In practice, the prefix 'mono' is usually omitted, except in the case of carbon monoxide. Examples are as follows:

N_2O dinitrogen oxide   CO   carbon monoxide
NO_2 nitrogen dioxide CO_2 carbon dioxide

Many hydrides of non-metals have well established trivial names, and

IUPAC recommends their continued use. Well known exampl   er, H_2O, ammonia, NH_3, and methane, CH_4.

Acids

Acids have long been recognised as sour-tasting substances which, in the presence of water, possess the property of changing the colour of litmus from blue to red. (Litmus is a dye obtained from certain lichens.) Molecules of all acids contain one or more atoms of hydrogen which can be replaced (either directly or indirectly) by atoms of metals to give compounds known as *salts*. **Acids are always named after the salts to which they give rise.**

One well known acid is the binary compound of formula HCl. Replacement of the hydrogen atom by metal atoms gives salts such as sodium chloride, NaCl, and calcium chloride, $CaCl_2$. In consequence, HCl is called hydrogen chloride. Hydrogen chloride is a gas, and gases in general are inconvenient substances to handle. Consequently, hydrogen chloride is normally used as an aqueous solution (*i.e.* a solution in water) known as hydrochloric acid. If the hydrogen chloride content is low, we talk about 'dilute hydrochloric acid'; if the hydrogen chloride content is high, we call the material 'concentrated hydrochloric acid'. Other acidic binary compounds are shown in Table 5.6.

Table 5.6 Nomenclature of acidic binary compounds

Formula	Name of salts	Name of pure binary compound	Name of aqueous solution
HF	Fluorides	Hydrogen fluoride	Hydrofluoric acid
HBr	Bromides	Hydrogen bromide	Hydrobromic acid
HI	Iodides	Hydrogen iodide	Hydriodic acid
H_2S	Sulphides	Hydrogen sulphide	—

Relatively few acids are binary compounds. Most are described as *oxoacids*, which means that their molecules contain combined oxygen as well as combined hydrogen. Their names end in either '-ous acid' or '-ic acid', depending on whether their salts are called '-ites' or '-ates'. For example, the salts of HNO_2 are called nit*rites*; consequently HNO_2 is known as nit*rous* acid. But the salts of HNO_3 are called nit*rates*, and the name of HNO_3 is nit*ric* acid. Other examples are as follows.

Formula	Name of salts	Name of acid
H_2SO_3	Sulphites	Sulphurous acid
H_2SO_4	Sulphates	Sulphuric acid

Aqueous solutions of oxoacids have the same names as the pure compounds.

It can be seen from the above examples that wherever an element gives rise to two oxoacids, it is the one with the lower oxygen content which is named as the '-ous acid'. The higher acid, *i.e.* the one of higher oxygen content, is always called the '-ic acid'. Compare, for instance, HNO_2 with HNO_3, and H_2SO_3 with H_2SO_4.

Bases

Acids can be *neutralised*, *i.e.* their acidity can be destroyed, in chemical reactions with compounds called *bases*. Some bases are electrovalent (*i.e.* ionic), while others are covalent (*i.e.* molecular).

Electrovalent bases comprise metal oxides and hydroxides, *e.g.* calcium oxide, CaO, and sodium hydroxide, NaOH. The best known covalent base is ammonia, NH_3. Ammonia is a gas, and, as with hydrogen chloride, we often prefer to use the substance in water. The solution is usually called aqueous ammonia, and given the formula $NH_3(aq)$.

Bases which are soluble in water are known as *alkalis*. The resultant solutions will turn litmus from red to blue and are said to be *basic* or *alkaline*.

Salts

Salts are the neutralisation products of acids and bases. For example, sodium chloride is the salt formed when hydrochloric acid is neutralised by the base, sodium hydroxide. Virtually all salts are electrovalent compounds and are named after the ions they contain.

Some salts are binary compounds of metals and non-metals and are named according to the rules stated above. Other salts are treated in a similar manner. In every case the positively charged ion is named first, followed by the negatively charged ion, and where necessary the valency of a metal is shown by a Roman numeral, *e.g.*

NaBr	sodium bromide	$NaNO_2$	sodium nitrite
ZnS	zinc sulphide	KCN	potassium cyanide
$MgSO_4$	magnesium sulphate	$Fe(NO_3)_2$	iron(II) nitrate
$(NH_4)_2CO_3$	ammonium carbonate	$Fe(NO_3)_3$	iron(III) nitrate

Organic compounds

There are so many compounds of carbon because the covalent bonds between carbon atoms are particularly strong. As a result, carbon atoms can join together in long chains, thus:

The chains can assume a variety of shapes, *e.g.*

Chains can branch, *e.g.*

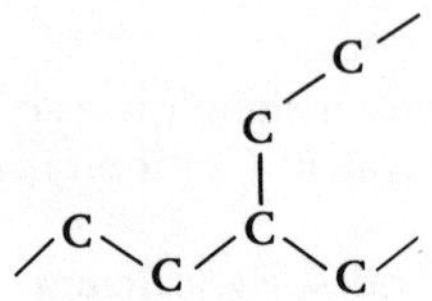

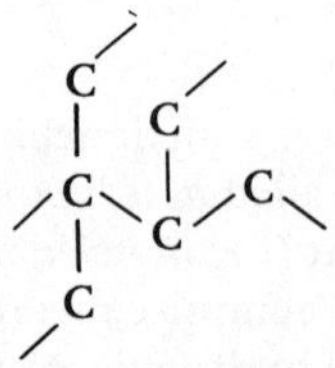

Two or more chains can join together or *cross-link, e.g.*

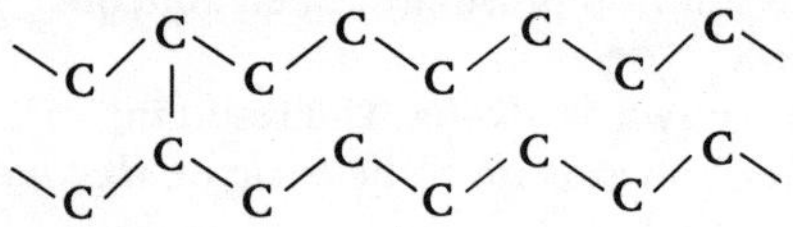

The ends of a chain can come together to form a 'ring', *e.g.*

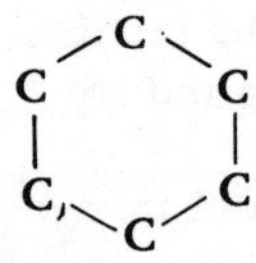

Two or more rings can join together, *e.g.*

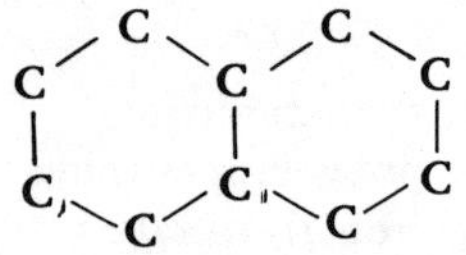

The possibilities are limitless. No other element has the ability to form chains and rings on such a scale.

We must not forget that carbon always exhibits a covalency of 4. The question therefore arises: 'To what other atoms are the carbon atoms attached?' The answer is that they may be joined to atoms of hydrogen, oxygen, nitrogen, sulphur, chlorine, bromine or iodine, but to little else. In general, they are mostly joined to atoms of hydrogen, thus:

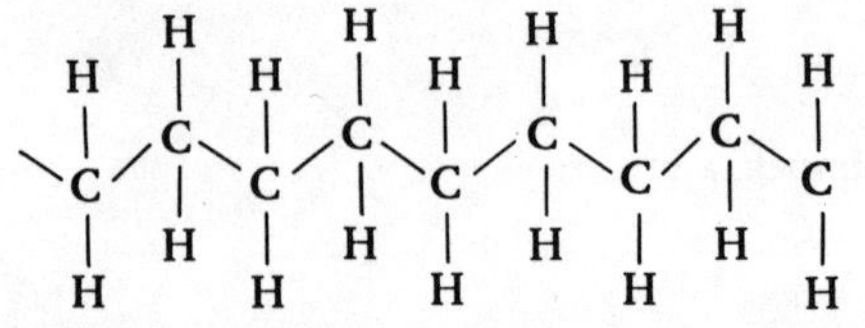

Chains of this kind are known as *hydrocarbon chains,* because they are composed of hydrogen and carbon. Some organic compounds consist of hydrogen and carbon only, in which case they are called *hydrocarbons.* Methane is a hydrocarbon.

The hydrocarbon chain shown above has been printed so as to show its shape, but chains are usually written with all the carbon atoms in a straight line, thus:

$$-\overset{\displaystyle H}{\underset{\displaystyle H}{\overset{|}{\underset{|}{C}}}}-\overset{\displaystyle H}{\underset{\displaystyle H}{\overset{|}{\underset{|}{C}}}}-\overset{\displaystyle H}{\underset{\displaystyle H}{\overset{|}{\underset{|}{C}}}}-\overset{\displaystyle H}{\underset{\displaystyle H}{\overset{|}{\underset{|}{C}}}}-\overset{\displaystyle H}{\underset{\displaystyle H}{\overset{|}{\underset{|}{C}}}}-\overset{\displaystyle H}{\underset{\displaystyle H}{\overset{|}{\underset{|}{C}}}}-\overset{\displaystyle H}{\underset{\displaystyle H}{\overset{|}{\underset{|}{C}}}}-\overset{\displaystyle H}{\underset{\displaystyle H}{\overset{|}{\underset{|}{C}}}}-\overset{\displaystyle H}{\underset{\displaystyle H}{\overset{|}{\underset{|}{C}}}}-$$

For this reason, such chains are often called 'straight chains'. Abbreviations for straight chains are permissible. We can write:

$$- CH_2 CH_2 CH_2 CH_2 CH_2 CH_2 CH_2 CH_2 CH_2 -$$

or $\quad (-CH_2 -)_n$

Organic compounds other than hydrocarbons possess at least one *functional group*, *i.e.* an atom or group of atoms that very largely controls the behaviour of the compound in chemical reactions. A typical functional group is −OH, the hydroxyl group, which is present in ethanol and indeed all alcohols.

$$H-\overset{\displaystyle H}{\underset{\displaystyle H}{\overset{|}{\underset{|}{C}}}}-\overset{\displaystyle H}{\underset{\displaystyle H}{\overset{|}{\underset{|}{C}}}}-O-H \qquad \text{ethanol}$$

In general terms, a simple organic molecule can be represented as follows:

The zig-zag line represents the hydrocarbon chain of the molecule, and X stands for the functional group.

Organic compounds are classified on the basis of functional group as alcohols, aldehydes, carboxylic acids, etc. Altogether, there are about a dozen functional groups, and hence about a dozen types of organic compound. Some of the commonest are shown in Table 5.7.

Alkanes

The alkanes are just one of several kinds of hydrocarbons recognised by organic chemists. They are distinguished by the fact that in their molecules *all* the covalent bonds between carbon atoms are single bonds, whereas in other types of hydrocarbons there is at least one multiple bond in the molecule. Alkanes have the general formula $C_n H_{2n+2}$, where n can be any number from 1 upwards.

Table 5.7 Classification of organic compounds on the basis of functional group

Functional group		Compounds which possess the group	Examples
Formula	Name		
$-OH$	Hydroxyl	Alcohols	Ethanol, CH_3CH_2OH
		Phenols	Phenol, C_6H_5OH
$\diagdown C=O \diagup$	Carbonyl	Aldehydes	Ethanal, $CH_3C\overset{\diagup O}{\diagdown H}$
		Ketones	Propanone, $CH_3\overset{O}{\overset{\|}{C}}CH_3$
$-C\overset{\diagup O}{\diagdown OH}$	Carboxyl	Carboxylic acids	Ethanoic acid, $CH_3C\overset{\diagup O}{\diagdown OH}$
$-Cl$	Chloro	Chlorides	Chloromethane, CH_3Cl

Table 5.8 Nomenclature of alkanes

Formula	Name	Formula	Name
CH_4	Methane	$CH_3(CH_2)_3CH_3$	Pentane
CH_3CH_3	Ethane	$CH_3(CH_2)_4CH_3$	Hexane
$CH_3CH_2CH_3$	Propane	$CH_3(CH_2)_5CH_3$	Heptane
$CH_3(CH_2)_2CH_3$	Butane	$CH_3(CH_2)_6CH_3$	Octane

Unbranched chain alkanes The formulae and names of the first eight unbranched alkanes are shown in Table 5.8.

Alkyl radicals In theory, an atom of hydrogen can be removed from any alkane molecule to give an *alkyl radical*. A radical is not a molecule; it is simply a group of atoms that commonly remains together and is encountered repeatedly in molecules of many different substances.

For example, the removal of a hydrogen atom from a molecule of methane leaves a methyl radical (commonly called a 'methyl group'), CH_3. Many organic molecules, *e.g.* CH_3OH (methanol) and CH_3COOH (ethanoic acid), contain a methyl radical.

From ethane there is derived the ethyl radical, CH_3CH_2, often written as C_2H_5.

Branched chain alkanes There is only one form of methane, *i.e.* there is only one way in which one carbon atom and four hydrogen atoms can be written so that carbon always has a valency of 4 and hydrogen a valency of 1. Likewise, there is only one form of ethane and one of propane. But there are two alkanes with the molecular formula C_4H_{10}. They are said to be *isomers* of each other, or *isomeric* with each other. 'Isomers', by definition, are compounds with the same molecular formula, but different properties.

The two isomers of butane are entirely distinct compounds. Each has its own structural formula, and consequently its own characteristic physical and chemical properties. Clearly, we cannot call them both 'butane', for this would invite confusion. According to the IUPAC rules for naming compounds systematically, the unbranched chain isomer is called butane, and the branched chain isomer methylpropane:

$$CH_3 CH_2 CH_2 CH_3 \qquad \begin{array}{c} CH_3 \\ \diagdown \\ CH_3 \diagup \end{array} CHCH_3$$

butane   methylpropane

To work out the systematic name of an alkane, we first look for the longest unbranched hydrocarbon chain in the molecule. We can, if we so wish, pencil a box around it:

$$\begin{array}{c} CH_3 \\ CH_3 \end{array} CHCH_3$$

This particular chain possesses three carbon atoms, so that the compound is regarded essentially as a derivative of propane.

Finally, we complete the name by writing down as a prefix the names of any alkyl radicals which are joined to the longest chain. In this case there is only one, namely a methyl radical, so that the full systematic name is methylpropane.

There are three isomers of pentane, $C_5 H_{12}$, as follows:

$$CH_3 CH_2 CH_2 CH_2 CH_3 \qquad \begin{array}{c} CH_3 \\ \diagdown \\ CH_3 \diagup \end{array} CHCH_2 CH_3 \qquad \begin{array}{c} CH_3 \\ | \\ CH_3 -C-CH_3 \\ | \\ CH_3 \end{array}$$

pentane   methylbutane   dimethylpropane

From this point onwards the number of possible isomers increases rapidly. There are five isomers of hexane, $C_6 H_{14}$, nine isomers of heptane, $C_7 H_{16}$, and 18 isomers of octane, $C_8 H_{18}$.

Source of Alkanes The lower alkanes occur as 'natural gas'. The principal source, for the UK, are the reserves underneath the North Sea, but natural gas is also found in America, Russia, the Middle East, Algeria, Holland, Norway and elsewhere. The composition of natural gas varies according to its origin, but in general the gas consists mainly of methane, with smaller amounts of ethane, propane, butane and nitrogen.

Alkanes from C_1 (methane) up to approximately C_{40} occur in the form of 'crude oil'. The principal deposits lie in the Middle East and Mexico, but other major oil producing areas include North America, Russia, North Africa and the UK. Crude oil, like natural gas, varies in composition according to its source. For example, oil from the North Sea is rich in alkanes of low relative molecular mass, but Middle Eastern crude contains a relatively high proportion of the heavier alkanes. Varying amounts of other compounds are also present, notably other sorts of hydrocarbons, acidic compounds, nitrogen containing compounds and sulphur containing compounds.

The first treatment that crude oil receives at a refinery is fractional distillation (see p. 90), which separates it into a number of fractions or 'cuts'; each fraction consists of a mixture of alkanes of approximately the same boiling temperature. The fractions taken, and their boiling ranges, vary from one refinery to another, but the general scheme is as follows:

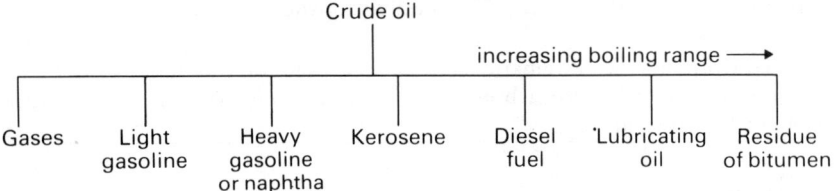

Gases

The gases, principally propane and butane, are readily liquefied by applying pressure at ordinary temperatures. In this form they are sold as LPG (liquefied petroleum gas) for industrial and domestic heating purposes. Some is converted to 'carbon black', a pure form of soot, by combustion in a limited supply of air. Carbon black is used as a pigment for paints and plastics, and as a filler for tyre rubber.

Light gasoline

After treatment to remove or modify the sulphur compounds that it contains, this fraction is suitable for use in motor cars. It is referred to as 'straight run petrol'.

Heavy gasoline or naphtha

Too volatile to be used with safety in jet aircraft, and too involatile to burn properly in motor car engines, this fraction is mostly *cracked* by subjecting it to heat, usually in the presence of a catalyst. Under these conditions an alkane molecule is broken down into two simpler hydrocarbon molecules.

Some heavy gasoline is not cracked but is refined and sold as 'white spirit', which is used as a paint thinner and also in dry cleaning.

Kerosene

With relatively little refining to remove sulphur compounds, this fraction is suitable as a fuel for jet aircraft. Some is sold as 'paraffin' for use in domestic heaters and oil-lamps.

Diesel fuel

The diesel fuel fraction is also little changed, the principal treatment again being the removal of sulphur compounds.

Lubricating oil

This is usually distilled into three fractions: light, medium and heavy. Light oil, which has the lowest boiling range, is the least viscous and is suitable for light machinery, such as sewing machines. Heavy machinery needs heavier, more viscous oils.

Bitumen

The tarry residue from the still is used partly in road making and partly in the manufacture of 'fuel oil' for industrial and domestic heating systems. For the latter purpose the viscosity of the bitumen must be reduced, *i.e.* it must be made to flow more easily. This can be achieved either by cracking the bitumen or by blending it with diesel fuel.

Summary

At the conclusion of this chapter, you should be able to:
1. state that compounds are of two main types, namely covalent (*i.e.* molecular) or electrovalent (*i.e.* ionic),
2. describe how atoms in molecules form covalent bonds,
3. recognise that covalent substances exist as simple molecules or giant molecules,
4. distinguish between empirical formulae, molecular formulae and structural formulae,
5. describe how ionic compounds are formed through electron transfer,
6. draw 'dot and cross' diagrams for binary covalent and electrovalent compounds,
7. recognise the stability of noble gas configurations as contributing to the stability of some electrovalent and covalent compounds,
8. state the formulae of molecules and ions, given their names,
9. classify substances as 'inorganic' or 'organic',
10. write the names of binary compounds, acids, bases and salts, given their formulae,
11. state that carbon atoms have the ability to form long chains in compounds,
12. state that carbon always has a covalency of 4,
13. define a 'hydrocarbon',
14. write the molecular and structural formulae of the lower alkanes,
15. define 'isomers' as substances which have the same molecular formula but different structures,
16. draw the molecular structures of simple isomeric alkanes,
17. write the systematic names of simple alkanes,
18. state the importance of oil as a source of hydrocarbons,
19. state the nature of the products of fractional distillation of crude oil.

Questions

In questions 1–10 select the most appropriate answer, labelled A, B, C or D.
1. In the formation of hydrogen chloride gas from hydrogen and chlorine,
   A H^+ and Cl^- ions are produced,
   B   hydrogen chloride molecules are produced, each of which consists of one molecule of hydrogen and one molecule of chlorine,

C   the bonding electron of a hydrogen atom stays put, while that of a chlorine atom is transferred to the hydrogen atom,

D   the nuclei of one hydrogen atom and one chlorine atom become mutually attracted to a shared pair of electrons.

2. When an atom of sodium combines with an atom of chlorine,

A   a pair of electrons is shared between the two atoms,

B   an electron is transferred from the sodium atom to the chlorine atom,

C   an electron is transferred from the chlorine atom to the sodium atom,

D   a molecule of sodium chloride is formed.

3. Which one of the following is *not* the electronic configuration of a noble gas?

A   2.

B   2, 8.

C   2, 8, 2.

D   2, 8, 8.

4. Tetrachloromethane (carbon tetrachloride), a compound of carbon and chlorine, is a liquid with a boiling temperature of 76.8 °C (350.0 K). It is most likely composed of:

A   simple molecules,

B   giant molecules,

C   positively charged carbon ions and negatively charged chloride ions,

D   positively charged chlorine ions and negatively charged carbide ions.

5. Barium (electronic configuration 2, 8, 18, 18, 8, 2) combines with chlorine (2, 8, 7) to give barium chloride, a crystalline solid with a high melting temperature. The structure is best represented by:

A $Ba-Cl$

B $Cl-Ba-Cl$

C $Ba^+ Cl^-$

D $Ba^{2+} (Cl^-)_2$

6. Sodium sulphate, an electrovalent compound, has the formula $Na_2 SO_4$. Its structure is best represented by:

A $\begin{array}{l} Na-O \\ Na-O \end{array} S \begin{array}{l} O \\ O \end{array}$

B $Na-O-O-S-O-O-Na$,

C $(Na^+)_2 SO_4{}^{2-}$,

D $(Na^+)_2 S^{6+} (O^{2-})_4$.

7. Which one of the following molecular formulae represents an alkane?

A $C_2 H_5$

B $C_3 H_9$

C $C_4 H_{10}$

D $C_5 H_{14}$

8. Which one of the following correctly represents the structural formula of an alkane?

$$\begin{array}{c} CH_3 \\ | \\ \end{array}$$

A $CH_3 CH_2 CHCH_3$

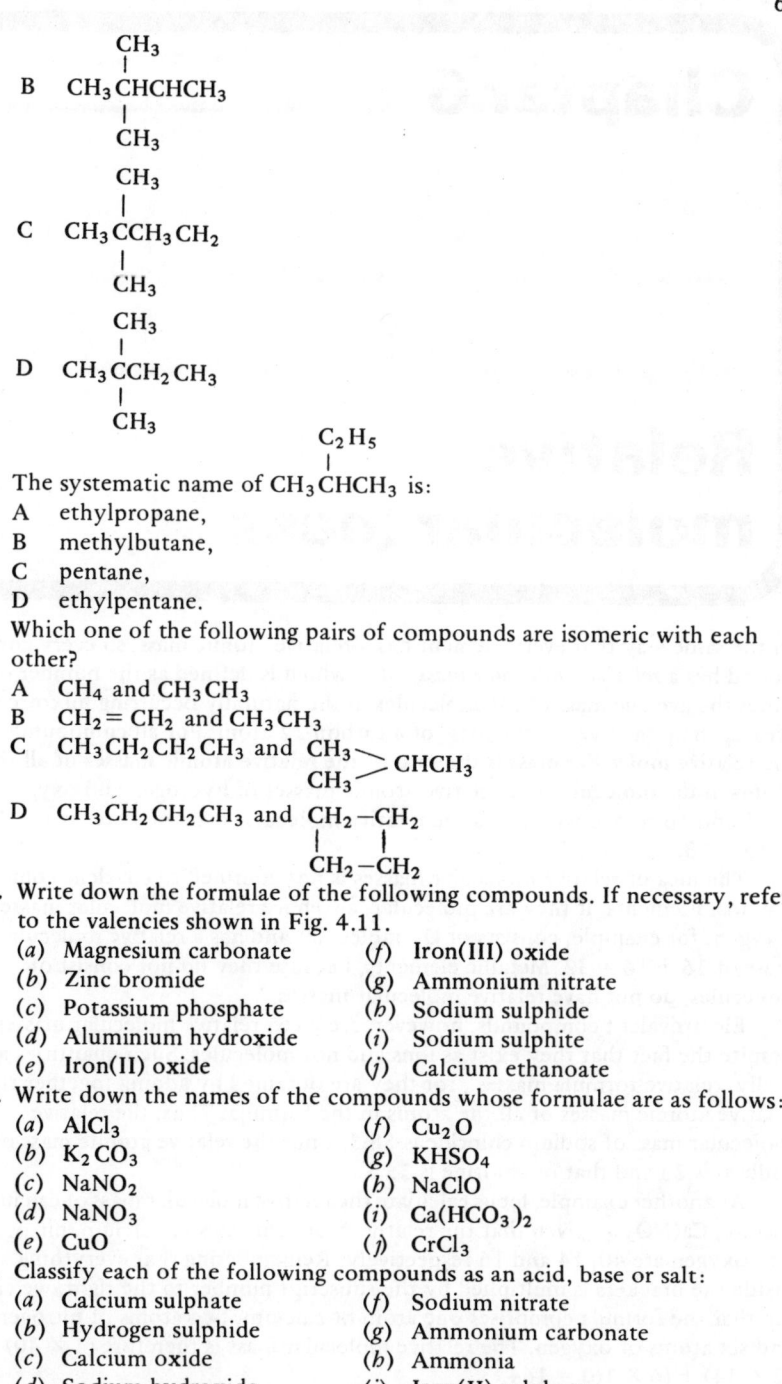

B   CH$_3$CHCHCH$_3$
with CH$_3$ substituent above and CH$_3$ below

C   CH$_3$CCH$_3$CH$_2$
with CH$_3$ substituent above and CH$_3$ below

D   CH$_3$CCH$_2$CH$_3$
with CH$_3$ substituent above and CH$_3$ below

9. The systematic name of CH$_3$CHCH$_3$ (with C$_2$H$_5$ substituent) is:
   A ethylpropane,
   B methylbutane,
   C pentane,
   D ethylpentane.

10. Which one of the following pairs of compounds are isomeric with each other?
   A CH$_4$ and CH$_3$CH$_3$
   B CH$_2$=CH$_2$ and CH$_3$CH$_3$
   C CH$_3$CH$_2$CH$_2$CH$_3$ and CH$_3$ / CH$_3$ CHCH$_3$
   D CH$_3$CH$_2$CH$_2$CH$_3$ and CH$_2$—CH$_2$ / CH$_2$—CH$_2$

11. Write down the formulae of the following compounds. If necessary, refer to the valencies shown in Fig. 4.11.
   (a) Magnesium carbonate   (f) Iron(III) oxide
   (b) Zinc bromide   (g) Ammonium nitrate
   (c) Potassium phosphate   (h) Sodium sulphide
   (d) Aluminium hydroxide   (i) Sodium sulphite
   (e) Iron(II) oxide   (j) Calcium ethanoate

12. Write down the names of the compounds whose formulae are as follows:
   (a) AlCl$_3$ (f) Cu$_2$O
   (b) K$_2$CO$_3$ (g) KHSO$_4$
   (c) NaNO$_2$ (h) NaClO
   (d) NaNO$_3$ (i) Ca(HCO$_3$)$_2$
   (e) CuO   (j) CrCl$_3$

13. Classify each of the following compounds as an acid, base or salt:
   (a) Calcium sulphate   (f) Sodium nitrate
   (b) Hydrogen sulphide   (g) Ammonium carbonate
   (c) Calcium oxide   (h) Ammonia
   (d) Sodium hydroxide   (i) Iron(II) sulphate
   (e) Hydrogen iodide   (j) Iron(III) sulphate

Chapter 6

Relative molecular mass

In the same way that every element has a relative atomic mass, so every compound has a *relative molecular mass* (M_r), which is defined as the number of times the average mass of the molecules in the naturally occurring mixture is greater than one-twelfth the mass of a carbon-12 atom. **For all compounds the relative molecular mass is the sum of the relative atomic masses of all the atoms in the molecule.** The relative atomic masses of hydrogen and oxygen are 1 and 16 respectively; thus the relative molecular mass of water is $1 + 1 + 16 = 18$.

The idea of relative molecular masses is not confined to covalent compounds. Elements, if they are molecular, also have relative molecular masses. Oxygen, for example, consists of O_2 molecules and has a relative molecular mass of $16 + 16 = 32$. Metallic elements, because they do not consist of molecules, do not have relative molecular masses.

Electrovalent compounds, however, are given relative molecular masses despite the fact that they exist as ions and not molecules. Such quantities are really 'relative formula masses', for they are obtained by adding together the relative atomic masses of all the atoms in the formula. Thus, the relative molecular mass of sodium chloride is 58.5, since the relative atomic mass of sodium is 23 and that of chlorine is 35.5.

As another example, let us calculate the relative molecular mass of calcium nitrate, $Ca(NO_3)_2$, given that the relative atomic masses of calcium, nitrogen and oxygen are 40, 14 and 16 respectively. Remembering that everything inside the brackets is multiplied by the subscript number to the right, we can see that the formula comprises one atom of calcium, two atoms of nitrogen and six atoms of oxygen. The relative molecular mass is therefore $(1 \times 40) + (2 \times 14) + (6 \times 16) = 164$.

Relative molecular masses, like relative atomic masses, can be determined by experiment, but this is a subject which must be left for a higher level of study.

Some uses of relative molecular masses

Relative molecular masses are required in most chemical calculations. We shall see this particularly in Chapter 9, but one or two examples are given here and in Chapter 7. Relative molecular masses also help to explain some of the laws of chemical combination, notably the Law of Constant Composition and the Law of Multiple Proportions.

Calculation of the percentages by mass of elements in compounds

Calculations of this kind are straightforward, once we accept that the symbols of elements and the formulae of compounds have a quantitative significance. Suppose, for example, that we need to know the percentages of oxygen and hydrogen in water. The fraction of oxygen which is chemically combined in water is represented by O/H_2O. Multiplication by 100 converts the fraction to a percentage. Thus, if we substitute the relative atomic mass of oxygen (16) and the relative molecular mass of water (18), we see that:

$$\text{oxygen content} \quad = \frac{16}{18} \times 100 = 88.9 \text{ per cent}$$

therefore, hydrogen content $\quad = 100 - 88.9 = 11.1$ per cent

As a check, we can argue that the fraction of hydrogen which is chemically combined in water is represented by $2H/H_2O$. Notice particularly $2H$, because there are two atoms of hydrogen in every molecule of water. The relative atomic mass of hydrogen is 1,

therefore, hydrogen content $= \dfrac{2 \times 1}{18} \times 100 = 11.1$ per cent

Calculation of the percentage of water of crystallisation in salt hydrates

Although sodium chloride crystallises from solution in the *anhydrous* state, *i.e.* without any chemically combined water, many other salts crystallise with one or more molecules of *water of crystallisation.* An example is copper(II) sulphate, $CuSO_4$, which crystallises with five molecules of water of crystallisation. The substance is known as a *salt hydrate.* Its name is copper(II) sulphate pentahydrate, or copper(II) sulphate 5-water, and its formula is $CuSO_4 \cdot 5H_2O$. In appearance it is a blue crystalline solid. It looks absolutely dry, because the water molecules are not simply mixed with the copper(II) sulphate, but are chemically joined to it and form an essential part of the crystal structure.

The percentage by mass of water of crystallisation in a salt hydrate is readily determined by a calculation similar to those shown immediately above. Thus, for copper(II) sulphate pentahydrate,

$$\text{water content} = \frac{5H_2O}{CuSO_4 \cdot 5H_2O} \times 100 \text{ per cent}$$

As we have seen, water has a relative molecular mass of 18. The relative atomic masses of copper, sulphur and oxygen are 63.5, 32 and 16 respectively; hence the relative molecular mass of $CuSO_4 \cdot 5H_2O$ is $(63.5 + 32 + 64) + 5(18) = 249.5$.

Therefore, water content $= \dfrac{5 \times 18}{249.5} \times 100 = 36.1$ per cent

This result can be verified in the laboratory by heating a known mass of copper(II) sulphate pentahydrate in a crucible supported by a pipeclay triangle resting on a tripod. Water of crystallisation is driven off, and anhydrous copper(II) sulphate, $CuSO_4$, remains behind as a white powdery solid. Heating is continued until the crucible and its contents have reached a constant mass. Subtraction of the final mass from the original mass gives the mass of water of crystallisation.

$$\text{Water content} = \frac{\text{mass of water of crystallisation}}{\text{mass of copper(II) sulphate pentahydrate}} \times 100 \text{ per cent}$$

Law of Constant Composition

This law, which is attributed to Joseph Proust (1799), states that: 'In any compound, irrespective of its method of preparation, the elements are always present in the same proportions by mass.'

The law is true because:

i.   the atoms of the various elements that make up a compound are present in a fixed ratio, regardless of how the compound is prepared,
ii. the relative atomic masses of elements and the relative molecular masses of compounds are fixed quantities.

In hydrogen chloride, for example, the ratio of hydrogen atoms to chlorine atoms is always 1:1. Hydrogen and chlorine have fixed relative atomic masses of 1 and 35.5 respectively, and hydrogen chloride has a fixed relative molecular mass of 36.5. Thus, in all samples of hydrogen chloride,

$$\text{proportion of hydrogen} = \frac{1}{36.5} = 0.027,$$

$$\text{and proportion of chlorine} = \frac{35.5}{36.5} = 0.973$$

Law of Multiple Proportions

This law was postulated by John Dalton (1802–4) and verified by Baron Berzelius in 1811 by a number of very accurate analyses. It reads as follows: 'If two elements combine together to form more than one compound, the various masses of the one element which combine with a fixed mass of the other are in a simple, whole number ratio.'

An example is provided by sulphur and oxygen, which combine together to give two oxides, namely sulphur dioxide and sulphur trioxide. If we select a fixed mass of one of these elements, say 1 g of sulphur, we can show by experiment that the masses of oxygen combining with this are as follows:

sulphur dioxide 1 g of sulphur combines with 1 g of oxygen,
sulphur trioxide 1 g of sulphur combines with 1.5 g of oxygen.

These amounts of oxygen are clearly in the ratio 2:3.

A similar result can be obtained by taking a fixed mass of oxygen and finding the various masses of sulphur which combine with it. For example,

sulphur dioxide 1 g of oxygen, by experiment, combines with 1 g of sulphur,
sulphur trioxide 1°g of oxygen combines with 2/3 g of sulphur.

The ratio of the masses of sulphur = 3:2, which again satisfies the Law of Multiple Proportions.

The law is true because:

i. atoms of different elements combine together in a simple, whole number ratio,
ii. the relative atomic masses of elements and the relative molecular masses of compounds are fixed quantities.

Suppose that we have two elements, A and B, giving rise to a series of compounds:

AB AB_2 AB_3 AB_4

Let the relative atomic masses of A and B be x and y respectively. Then the relative molecular masses of the compounds will be as follows:

	AB	AB_2	AB_3	AB_4
relative molecular mass	$x + y$	$x + 2y$	$x + 3y$	$x + 4y$

In each case the mass of A is fixed, and the masses of B are in the ratio 1:2:3:4.

The mole concept

When working in the laboratory, we cannot study individual molecules of a substance for they are far too small. We can work only with a very large number of molecules. In other words, we have to scale everything up to such a level that we can readily observe what is happening.

Scaling up a chemical reaction is rather like scaling up a jam recipe, for in both cases we have to multiply by a suitable factor. For example, we could, in theory, make jam by heating together 1 gram of fruit with 2 grams of sugar, but it would be very difficult to work with these small quantities. The recipe needs to be scaled up. We could multiply by a factor of two, and take 2 g of fruit and 4 g of sugar, or by a factor of, say, five and take 5 g of fruit and 10 g of sugar. But the neatest way of scaling things up is simply to change the units. Instead of using 1 gram of fruit and 2 grams of sugar, we could take 1 *kilogram* of fruit and 2 *kilograms* of sugar. By changing the mass units from grams to kilograms we are in effect multiplying the recipe by a factor of a thousand.

For domestic purposes this would be a convenient jam recipe, but for the food industry the batch size would still be too small. A greater degree of scaling is needed, and industry might well alter the units from grams to tonnes. (The recipe now becomes 1 tonne of fruit and 2 tonnes of sugar.) The change from grams to tonnes represents a scaling up of the basic recipe by a factor of a million.

Let us now switch from a jam recipe involving 1 gram of fruit and 2 grams of sugar to a chemical reaction in which 1 molecule of sulphuric acid (H_2SO_4) reacts with 2 molecules of ammonia (NH_3). Once more we can scale things up, and keep the 1:2 ratio intact, by changing the mass units.

But what are the mass units of molecules? A molecule, as we know, is given a *relative* mass, which is the number of times its mass is greater than 1/12th the mass of a carbon-12 atom. However, we can, if we choose, regard 1/12th the mass of a carbon-12 atom as a unit of mass. Let us call it the 'atomic mass unit' (u). With this rather unconventional unit we can write down molecular masses. We can say, for example, that sulphuric acid, which has a relative molecular mass of 98, has a molecular mass of 98 u. Likewise, ammonia has a relative molecular mass of 17 and a molecular mass of 17 u.

If 1 molecule of sulphuric acid reacts with two molecules of ammonia, then in terms of mass 98 u of sulphuric acid react with $2 \times 17 = 34$ u of ammonia. We can now scale up to a practical level by altering 'atomic mass units' (u) to grams (g). Thus, we can say that 98 g of sulphuric acid react with 34 g of ammonia.

The amount of any substance represented by its relative molecular mass expressed in grams is said to be one mole of that substance. (For metallic elements, and also for certain non-metals, such as carbon, which exist as giant molecules, a mole is the relative *atomic* mass expressed in grams.) Thus, 98 g of sulphuric acid represents 1 mol of sulphuric acid. In the case of ammonia, 17 g is 1 mol and 34 g is 2 mol. ('mol' is the accepted abbreviation for mole and moles.)

We know that

1 molecule of H_2SO_4 reacts with 2 molecules of NH_3,
i.e. 98 u of H_2SO_4 reacts with 34 u of NH_3

Therefore, after changing the units,

98 g of H_2SO_4 reacts with 34 g of NH_3,
i.e. 1 mol of H_2SO_4 reacts with 2 mol of NH_3

Notice particularly that the reacting ratio of sulphuric acid to ammonia is the same on the molar scale as it is on the molecular scale. **It is always true that the ratio in which moles of substances react together is the same as that in which their molecules react together.**

There is no obligation to work on a one molar scale, but we must always aim to work on some fraction or multiple of the molar scale. For example, the reaction between sulphuric acid and ammonia could be conducted on a tenth molar scale, in which case we should need 9.8 g of sulphuric acid and 3.4 g of ammonia; or we could work on, say, a twice molar scale with 196 g of sulphuric acid and 68 g of ammonia.

Conversion of grams to moles and vice versa

These are vital yet very simple calculations. We shall explain them with reference to water, but the principles apply equally to all other substances.

Water has a relative molecular mass of 18; therefore 18 g represents one mole of water.

If 18 g of water is 1 mol,

then 1 g of water is $\dfrac{1}{18}$ mol,

and x g of water is $\dfrac{1}{18} \times x$ mol.

To summarise, **we convert grams to moles by dividing by relative molecular mass.**

If 1 mol of water is 18 g,
then x mol of water is $18 \times x$ g.
To summarise, **we convert moles to grams by multiplying by relative molecular mass.**

The Avogadro constant

We saw earlier in this chapter that the change from grams to kilograms corresponds to multiplication by a thousand, *i.e.* the number of grams in a kilogram. Likewise, the change from grams to tonnes corresponds to multiplication by a million, *i.e.* the number of grams in a tonne. In the same way, the change from 'atomic mass units' to grams corresponds to multiplication by a factor which must be equal to the number of 'atomic mass units' in a gram. This proportionality factor is known as the *Avogadro constant* (L).

We are not allowed to define the Avogadro constant in this way, because the 'atomic mass unit' is not a recognised unit of mass, but this does provide a starting point for deriving an acceptable definition.

If L is the number of 'atomic mass units' in 1 gram,
then L is the number of times that 1 g is greater than 1 'atomic mass unit',
i.e. " " " " " " " " "1/12th the mass of a carbon-12 atom,
therefore " " " " 12g " " "the entire mass of a carbon-12 atom,
therefore mass of one carbon-12 atom $\times L = 12$ g,
therefore mass of L carbon-12 atoms $= 12$ g.
Thus, L **is the number of atoms in 12 g of carbon-12.** This is the generally accepted definition of the Avogadro constant, L.

Twelve grams of carbon-12 represent 1 mole of carbon-12. Consequently, we can say that L is the number of atoms in 1 mole of carbon-12. The same is true of substances other than carbon-12. Whenever we switch from a molecule (or atom, in the case of certain elements) to a mole, we multiply by a factor of L. Thus, there are always L molecules (or atoms, as the case may be) in a mole. **A mole can therefore be defined as the amount of substance containing the number of particles equal to the Avogadro constant.**

Estimation of the size and mass of a molecule, and the value of the Avogadro constant.

This method is based on the fact that when a small quantity of stearic acid, $CH_3(CH_2)_{16}COOH$, is added to water, the compound spreads out and forms a layer one molecule thick, known as a *monolayer*. If a loop of cotton thread is placed on the water, it will trap the stearic acid molecules and prevent them spreading across the entire surface (see Fig. 6.1).

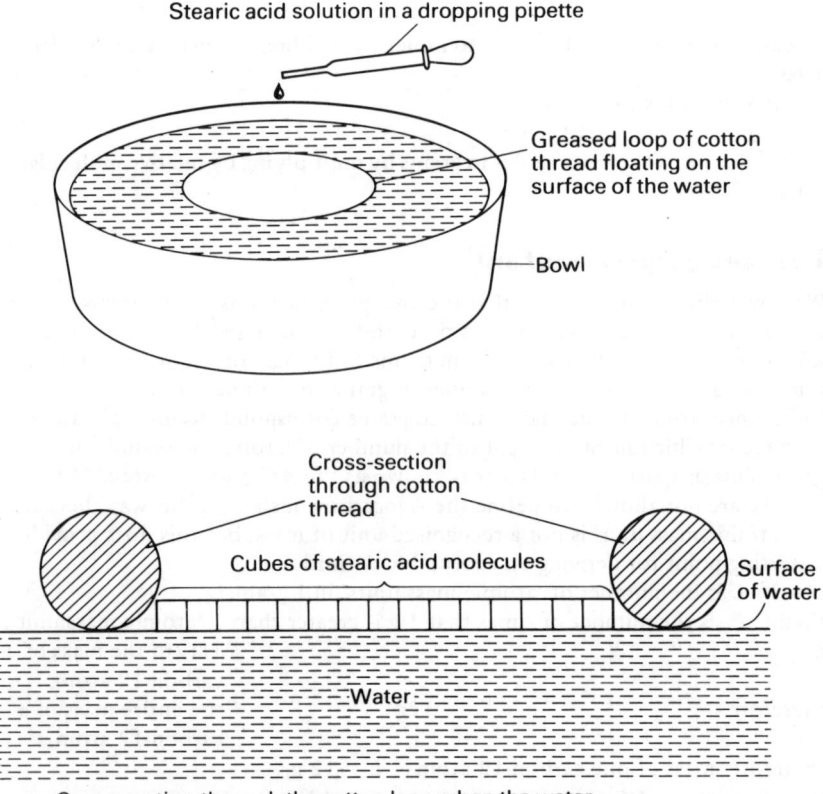

Stearic acid solution in a dropping pipette

Greased loop of cotton thread floating on the surface of the water

Bowl

Cross-section through cotton thread

Cubes of stearic acid molecules

Surface of water

Water

Cross-section through the cotton loop when the water is covered by a single layer of cubic stearic acid molecules. The stearic acid molecules are drawn overscale compared with the cotton.

Fig. 6.1 A monolayer of stearic acid, enclosed by a loop of cotton thread

The experiment requires the use of a solution of stearic acid (or similar compound, such as oleic acid), in petroleum ether. A concentration of 0.1 g dm^{-3} is recommended. Also needed are a bowl, about 25 cm in diameter, filled with water, a lightly greased loop of thread with a circumference of about 30 cm, a dropping pipette and a 10 cm^3 measuring cylinder.

Instructions for performing the experiment are as follows.

1. Float the loop of thread on the surface of the water. Hold the dropping pipette containing the stearic acid solution over the loop, squeeze the bulb gently, and allow drops of solution to fall inside the loop. Count the minimum number of drops required to make the loop moderately rigid, so that it can be moved like a raft.

   The petroleum ether evaporates rapidly to leave only the stearic acid on the surface, and the loop becomes stiff as soon as it contains a monolayer of stearic acid.

2. Remove the loop of thread, and determine its circumference by cutting it open and laying it on a ruler. (Alternatively, pull the loop taut, lay the double thickness of thread on a ruler, and obtain the circumference of the loop by multiplying the length by two.)

3. By means of the measuring cylinder, find the number of drops of stearic acid solution there are in 1 cm^3. Measuring cylinders are notoriously inaccurate at the bottom, so it is advisable, first, to pour some solution into the cylinder, and then count the number of drops of solution required to increase the volume by 1 cm^3.

Specimen results

The following is a set of results obtained by a student.

i. Number of drops of solution required to make the loop rigid = 5

ii. Circumference of the loop = 31.0 cm

iii. Number of drops of solution in 1 cm^3 = 70

Calculation of the Avogadro constant

L is found by a calculation in the course of which we can obtain values for both molecular size and molecular mass. Instructions for each stage are given below and illustrated by the specimen results quoted above.

1. Calculate the area of the loop

   Circumference of a circle = $2\pi r$ (r = radius)

   therefore $\quad\quad 31.0 = 2 \times \dfrac{22}{7} \times r$

   therefore $\quad\quad r = \dfrac{31.0 \times 7}{2 \times 22} = 4.93\,cm$

   Area of a circle = πr^2

   $\quad\quad\quad\quad = \dfrac{22}{7} \times 4.93^2$

   $\quad\quad\quad\quad = 76.4\ cm^2$

2. Calculate the volume of stearic acid solution in the loop
   70 drops of solution have a volume of 1 cm^3,

   therefore   1 drop of solution has a volume of $\dfrac{1}{70}$ cm^3,

   and   5 drops of solution have a volume of $\dfrac{1}{70} \times 5 = 0.0714$ cm^3

3. Calculate the mass of stearic acid in the loop
   1000 cm^3 of solution contain 0.1 g of stearic acid,

   therefore 1 cm^3 of solution contains $\dfrac{0.1}{1000}$ g of stearic acid,

   and 0.0714 cm^3 of solution contains $\dfrac{0.1}{1000} \times 0.0714$ g

   $$= 0.000\ 00714\ \text{g}$$
   $$= 7.14 \times 10^{-6}\ \text{g of stearic acid}$$

4. Calculate the volume of stearic acid in the loop
   The density of stearic acid is known to be 0.941 g cm^{-3}

   $$\text{density} = \frac{\text{mass}}{\text{volume}},$$

   therefore $\text{volume} = \dfrac{\text{mass}}{\text{density}}$

   $$= \frac{7.14 \times 10^{-6}}{0.941} = 7.59 \times 10^{-6}\ \text{cm}^3$$

5. Calculate the thickness of the monolayer
   The volume of the layer = area $\times$ thickness,

   therefore thickness $= \dfrac{\text{volume}}{\text{area}}$

   $$= \frac{7.59 \times 10^{-6}}{76.4}$$

   $$= 0.0993 \times 10^{-6}$$
   $$= 9.93 \times 10^{-8}\ \text{cm}$$

6. Calculate the volume of one molecule of stearic acid.
   For this purpose we assume that the stearic acid molecules are cubes. The sides of any cube have the same length (l), and the volume of a cube is l^3.
   The length of each side of a 'stearic acid molecule cube' is equal to the thickness of the monolayer,

   therefore volume of one molecule $= (9.93 \times 10^{-8})^3 = 979 \times 10^{-24}$
   $$= 9.79 \times 10^{-22}\ \text{cm}^3$$

   If necessary, we can find the mass of a molecule at this stage by multiplying volume by density,

   therefore mass of a stearic acid molecule $= 9.79 \times 10^{-22} \times 0.941$
   $$= 9.21 \times 10^{-22}\ \text{g}$$

7. Calculate the volume of one mole of stearic acid.
   The relative molecular mass of stearic acid is 284; therefore one mole of

stearic acid is 284 g. We know that the density of stearic acid is 0.941 g cm^{-3}.

If 0.941 g occupies a volume of 1 cm^3,

then 1 g occupies a volume of $\dfrac{1}{0.941}$ cm^3,

and 284 g occupies a volume of $\dfrac{1}{0.941} \times 284 = 302$ cm^3 mol^{-1}

8. Calculate the Avogadro constant

L is equal to the number of times that the volume of one mole is greater than the volume of one molecule,

$$i.e.\ L = \frac{302}{9.79 \times 10^{-22}} = \frac{30.8}{10^{-22}} = 30.8 \times 10^{22}$$

$$= 3.08 \times 10^{23}\ mol^{-1}$$

The generally accepted value of the Avogadro constant is 6.02×10^{23} mol^{-1}

Summary

At the conclusion of this chapter, you should be able to:
1. define 'relative molecular mass',
2. calculate the relative molecular mass of a compound by the addition of relative atomic masses,
3. calculate the percentages by mass of elements in a compound,
4. calculate the percentage of water of crystallisation in a salt hydrate,
5. state the Law of Constant Composition,
6. state the Law of Multiple Proportions,
7. convert the mass of a substance to an amount in moles,
8. convert the amount of a substance in moles to a mass,
9. define the Avogadro constant as the number of atoms in 12 g of carbon-12,
10. define the mole as the amount of a substance containing the number of particles equal to the Avogadro constant,
11. estimate the size and mass of a molecule, and the value of the Avogadro constant, by means of a monolayer experiment.

Questions

1. Calculate the following. All the necessary relative atomic masses will be found in Fig. 4.2.
   (a) The relative molecular mass of nitric acid, HNO_3.
   (b) The relative molecular mass of sodium carbonate, $Na_2 CO_3$.
   (c) The percentages by mass of hydrogen, sulphur and oxygen in sulphuric acid, $H_2 SO_4$.
   (d) The percentage by mass of water of crystallisation in sodium carbonate decahydrate, $Na_2 CO_3 \cdot 10H_2 O$.
   (e) The mass (in grams) of two moles of nitrogen, N_2.
   (f) The mass (in grams) of half a mole of aluminium sulphate, $Al_2 (SO_4)_3$.

(g) The amount (in moles) represented by 100 g of sodium chloride.

(h) The amount (in moles) represented by 1 g of copper(II) sulphate pentahydrate, $CuSO_4 \cdot 5H_2O$.

2. Which of the following statements about the Avogadro constant (L) are true and which are false?

(a) It is the number of atoms in one molecule of carbon-12.

(b) It is the number of times that the volume of one mole of a substance is greater than the volume of one molecule.

(c) It is the number of molecules in one mole of any substance.

(d) It is the number of atoms in one gram of carbon-12.

(e) It is the number of atoms in twelve grams of carbon-12.

(f) It is the number of atoms in twelve moles of carbon-12.

(g) It is the number of times that one gram is greater than one-twelfth the mass of a carbon-12 atom.

(h) It is the number of times that twelve grams is greater than one-twelfth the mass of a carbon-12 atom.

Chapter 7

Solutions

Introduction

The formation of a solution

We all know that the majority of substances or mixtures of substances that we encounter in everyday life do not mix with water. Rocks of various kinds, most metals, plastics, wood, glass, cloth, etc., all remain unchanged in the presence of water. Granted, they become wet, but if a substance of this sort is thoroughly dried its mass is the same as the original, showing that none of it is lost to the water. Such substances are said to be *insoluble* in water.

By contrast, there are some substances, such as sugar or salt, which *dissolve, i.e.* break up, when immersed in water. Their crystals gradually become smaller and smaller and eventually disappear altogether.

A solid dissolves because the water penetrates between the individual molecules or ions of which the crystal is composed. As a result, the molecules or ions are detached from the crystal and become thoroughly mixed up with the water molecules. The separation of ions which occurs when an electrovalent compound dissolves is called *dissociation*.

Sugar is said to be *soluble* in water, and the homogeneous mixture which results is known as a *solution* (see Fig. 7.1(a)). Salt, also, is soluble in water to give a solution (see Fig. 7.1(b)). 'Homogeneous' means that the mixture has the same composition throughout, *i.e.* that the dissolved substance is evenly distributed throughout the solution. Individual particles of the dissolved substance cannot be seen, even with a high powered microscope, because they are single molecules or ions.

Some solutions have special names, *e.g.* 'brine' for salt in water, and

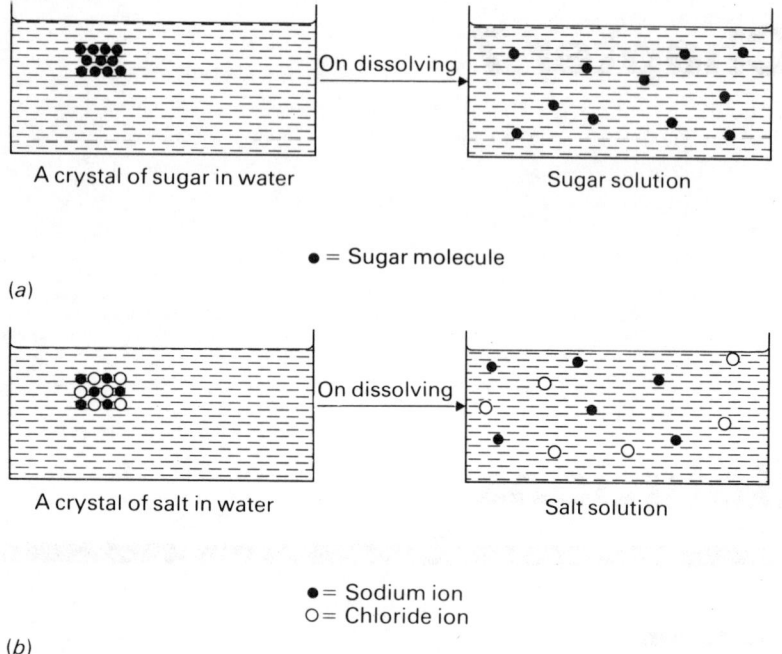

Fig. 7.1 The formation of aqueous solutions of (*a*) sugar, and (*b*) sodium chloride. The horizontal shading in all the diagrams represents large numbers of water molecules.

'syrup' for sugar in water, but most solutions have not. A solution of sodium hydroxide in water, for example, is referred to as 'aqueous sodium hydroxide', or simply 'sodium hydroxide solution'. The last term, although in common use, is not recommended because it is ambiguous. Solid sodium hydroxide will dissolve in other liquids besides water, and it is advisable, when referring to a solution, to specify both the dissolved substance and the liquid which does the dissolving. Descriptions such as 'aqueous sodium hydroxide' or 'alcoholic sodium hydroxide' avoid possible confusion.

There are special names given to both the substances which make up a solution. The dissolved substance (*e.g.* sugar, salt or sodium hydroxide) is called the *solute*, while the substance which does the dissolving is called the *solvent*. We can therefore define a solution as **a homogeneous mixture of a solute and a solvent.**

It is important not to confuse 'dissolving' with 'melting'. 'Melting' is the change of state of a single substance from solid to liquid brought about by a rise of temperature, while 'dissolving' necessitates the use of a solvent so as to form a solution. An increase of temperature simply speeds up the dissolving process.

Solutions and suspensions

Another distinction that must be made is that between a 'solution' and a 'suspension'. A *suspension* is formed whenever a finely divided solid is shaken with a liquid in which it is insoluble. For example, when chalk dust is shaken with water the particles of chalk become scattered in all directions, and if they are sufficiently small they will remain suspended in the water for a few minutes after the shaking has ceased. However, a suspension, unlike a solution, settles out after a time, and the liquid can be decanted from the sediment (see Fig. 7.2).

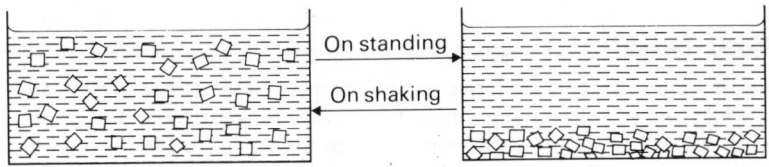

□ = Crystal of calcium carbonate

Fig. 7.2 A suspension of chalk dust (calcium carbonate) in water

In the formation of a suspension, the liquid never penetrates between the individual molecules or ions of which the solid is composed. All that happens is that the small crystals of solid, each of which contains many millions of molecules or ions, become distributed throughout the liquid.

The particles in a suspension are enormous by comparison with those of a solute in a solution, which, as we have seen, are single molecules or ions. Often, the particles in a suspension are visible to the naked eye; failing that, they will show up under a magnifying glass.

Types of solution

The only solutions which we have discussed so far are those of solids in liquids, but altogether six types can be made, as shown in Table 7.1. Other possible combinations of solid, liquid and gas do not give rise to solutions.

For all types of solutions the terms 'solute' and 'solvent' are used to describe the constituent substances. In many cases there is no doubt as to

Table 7.1 Types of solutions

Type of solution	Example
Solid in liquid	Brine (salt in water)
Solid in solid	An alloy of silver and gold
Liquid in liquid	Vodka (ethanol in water)
Gas in gas	Air (oxygen in nitrogen)
Gas in liquid	Soda water (carbon dioxide in water)
Gas in solid	Ammonia in charcoal

which is which. For instance, when carbon dioxide gas is dissolved in water to give 'soda water', carbon dioxide is the solute and water is the solvent. But when two gases or two liquids or two solids are dissolved in each other, the position is far less clear. Do we regard air as a solution of oxygen in nitrogen or nitrogen in oxygen? The answer is that the constituent which is present in the larger proportion is the solvent, the other being the solute. Thus, air, which consists of approximately 80 per cent of nitrogen and 20 per cent of oxygen, can be described as a solution of oxygen (the solute) in nitrogen (the solvent).

Concentration

A solution which contains a relatively small proportion of solute is said to be *dilute*, while one which contains a high proportion of solute is said to be *concentrated*. The amount of solute which is present in a certain volume of solution is known as the *concentration* of that solution.

The popular term for concentration is 'strength'. We often describe dilute solutions as 'weak' and concentrated solutions as 'strong'. However, the terms 'weak' and 'strong' have other meanings in chemistry and are best avoided here.

The concentration of a solution can be expressed in several ways, some of which are as follows.

Percentage by mass (mass/mass)

This can mean *either* the number of grams of solute in 100 grams of solvent, *or* the number of grams of solute in 100 grams of solution. If, therefore, we meet a solution in the laboratory labelled, say, '10 per cent potassium iodide', we cannot be certain of its precise concentration. The label could mean that the solution consists of:

i.   10 g of potassium iodide plus 100 g of water, or
ii.   10 g of potassium iodide plus 90 g of water.

Because of this ambiguity, concentration is seldom quoted these days on a mass/mass basis. Whenever the method is used (and we shall encounter it in connection with solubility later in this chapter) the units should be clearly stated, *e.g.* 10 g of solute per 100 g of water.

Mass concentration (mass/volume)

The phrase *mass concentration* refers to a concentration in terms of **mass of solute per unit volume of solution**. It is most conveniently expressed in units of $g\ dm^{-3}$, *i.e.* as the number of grams of solute in one cubic decimetre of solution. (The name 'litre' for the cubic decimetre is still widely used, and units of $g\ l^{-1}$ are acceptable.)

If basic SI units are preferred, the units for mass concentration are $kg\ m^{-3}$. The numerical value of a mass concentration is the same, regardless of whether we use units of $g\ dm^{-3}$ or $kg\ m^{-3}$. This is because there are 1000 grams in a kilogram, and also 1000 cubic decimetres in a cubic metre.

Concentration (molar mass/volume, 'molarity' or 'molar concentration')

According to the International Union of Pure and Applied Chemistry (IUPAC), the term 'concentration' should be defined in only one way, namely as **the amount of solute, in moles, per unit volume of solution.** The unit of volume most commonly used in chemistry is the cubic decimetre; therefore, concentration expressed in this way has units of mol dm^{-3}.

It will undoubtedly be many years until the term 'concentration' acquires this meaning only. In the meantime, although this is not recommended by IUPAC, the terms 'molarity' or 'molar concentration' may be used instead of 'concentration' to stress that the units are mol dm^{-3}. If a certain solution has a concentration of one mole per cubic decimetre, it can be said to be 'molar' or 'one molar', and its concentration can be written as M or 1 M instead of 1 mol dm^{-3}. Likewise, instead of 2 mol dm^{-3} or 0.1 mol dm^{-3}, we can write 2 M or 0.1 M respectively, and so on.

Most laboratory reagents have a concentration of 4 mol dm^{-3} (4 M). Throughout this book concentrations will be quoted, as here, in mol dm^{-3} with the corresponding molarity in brackets at the side.

Working in terms of concentration (molarity) can pose minor practical difficulties. For example, how much solute must we weigh out in order to prepare a solution of stated concentration (molarity)? How can we convert a concentration (molarity) to a mass concentration or vice versa? All these problems can be speedily resolved when it is recalled (see p. 71) that we can convert:

i. moles to grams by multiplying by relative molecular mass,
ii. grams to moles by dividing by relative molecular mass.

A few examples should make this clear.

Example 1

How many grams of anhydrous sodium carbonate, Na_2CO_3, must be weighed out in order to prepare 250 cm^3 of solution of concentration 0.1 mol dm^{-3} (0.1 M)? (Relative atomic masses: C = 12, O = 16, Na = 23)

Answer

The relative molecular mass of sodium carbonate is 106,
therefore 0.1 mol of Na_2CO_3 is 0.1 × 106 = 10.6 g
If 1000 cm^3 (*i.e.* 1 dm^3) of solution requires 10.6 g of Na_2CO_3,

then 1 cm^3 of solution requires $\dfrac{10.6}{1000}$ g of Na_2CO_3,

and 250 cm^3 of solution require $\dfrac{10.6}{1000}$ × 250 = 2.65 g of Na_2CO_3

Example 2

How many grams of copper(II) sulphate pentahydrate, $CuSO_4 \cdot 5H_2O$, are needed to make 5 dm^3 of solution of concentration 4 mol dm^{-3} (4 M)? (Relative atomic masses: H = 1, O = 16, S = 32, Cu = 63.5)

Answer

Water of crystallisation must always be taken into account in these calculations.
   The relative molecular mass of copper(II) sulphate pentahydrate is 249.5,
therefore 4 mol of $CuSO_4 \cdot 5H_2O$ is $4 \times 249.5 = 998$ g
If 1 dm^3 of solution requires 998 g of $CuSO_4.5H_2O$,
then 5 dm^3 of solution require $998 \times 5 = 4990$ g of $CuSO_4 \cdot 5H_2O$

Example 3

250 cm^3 of a salt solution contain 1.52 g of sodium chloride, NaCl. What is
its concentration (molarity)? (Relative atomic masses: Na $= 23$, Cl $= 35.5$)

Answer

The relative molecular mass of sodium chloride is 58.5.
Since there are   1.52 g of NaCl in 1/4 dm^3 (*i.e.* 250 cm^3) of solution,
there are 1.52×4 g of NaCl in 1 dm^3 of solution,
therefore there are $\dfrac{1.52 \times 4}{58.5} = 0.104$ mol in 1 dm^3 of solution (0.104 M).

Example 4

How many grams per cubic decimetre of sodium hydroxide, NaOH, are there
in a solution of concentration 0.102 mol dm^{-3} (0.102 M)? (Relative atomic
masses: H $= 1$, O $= 16$, Na $= 23$)

Answer

The relative molecular mass of sodium hydroxide is 40,
therefore 0.102 mol $dm^{-3} = 0.102 \times 40 = 4.08$ g dm^{-3} of NaOH

Example 5

How many grams per cubic decimetre of combined nitrogen are there in an
ammonia solution of concentration 0.098 mol dm^{-3} (0.098 M)? (Relative
atomic masses: H $= 1$, N $= 14$)

Answer

One molecule of ammonia, NH_3, contains one atom of nitrogen,
therefore 1 mol of NH_3 contains 1 mol of N,
therefore 0.098 mol dm^{-3} of NH_3 solution contains 0.098 mol dm^{-3} of N.
The relative atomic mass of nitrogen is 14,
therefore 0.098 mol dm^{-3} of N $= 0.098 \times 14 = 1.37$ g dm^{-3} of nitrogen

Preparation of a standard solution, i.e. a solution of stated concentration (molarity)

It is seldom necessary to prepare a solution to an exactly stated concentration
(molarity), *e.g.* 0.100 mol dm^{-3} (0.100 M). The task is by no means impossible,
but since it involves weighing out a precise amount of material it is a tricky and
time-consuming operation which should be attempted only when absolutely
necessary.

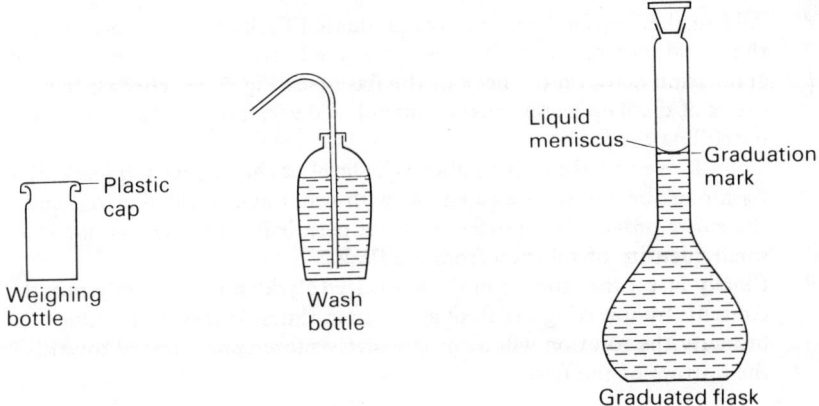

Fig. 7.3 Apparatus used in the preparation of a standard solution

For most purposes it is sufficient to prepare a solution which corresponds approximately to the stated concentration (molarity). Provided that the mass of the solute is accurately known, the exact concentration (molarity) of the solution can be calculated.

Experimental instructions are as follows. It is important that all glass apparatus is chemically clean, *i.e.* uncontaminated by unwanted substances. If there is any doubt, the apparatus should be thoroughly washed, first with detergent, then with tap water, and finally with distilled water. Weighing bottles must be both clean and dry.

1. Find the approximate mass of solute which is required by a calculation similar to those in examples 1 and 2 above.
2. Weigh an empty weighing bottle (see Fig. 7.3) on an analytical balance to the maximum accuracy permitted by the instrument, and record the weighing directly into your laboratory notebook; **not on a piece of scrap paper!**
3. By means of a spatula, transfer roughly the right amount of solute into the weighing bottle, reweigh, and again record the weighing in your notebook.
4. Find by subtraction the amount of solute that has been taken and calculate, as in example 3 above, the exact concentration (molarity) of the solution that will result.
5. Tip the contents of the weighing bottle into a beaker and rinse the weighing bottle with distilled water from a wash-bottle (see Fig. 7.3), making sure that all the washings enter the beaker.
6. Pour sufficient distilled water into the beaker to dissolve the solute, and stir with a glass rod until all the solute has dissolved. If necessary, heat the beaker on a tripod and gauze to assist dissolving.
7. Pour the cold solution into a graduated flask (see Fig. 7.3) of the appropriate size, taking care to avoid spillage. Wash the beaker and stirring rod at least twice with distilled water, and transfer all washings to the graduated flask. Mix the contents of the flask by swirling.

8. Add further distilled water to the graduated flask until the bottom of the liquid meniscus (*i.e.* the crescent shaped surface) coincides with the graduation mark on the neck of the flask (see Fig. 7.3). The last few drops of distilled water must be introduced very carefully so as to avoid overfilling the flask.

   Note If ever a flask is filled above the level of the graduation mark, the flask must be emptied, washed out with water and distilled water, and the entire procedure repeated. It is not permissible merely to remove a small quantity of solution from the flask.

9. Finally insert the stopper in the graduated flask and thoroughly mix the contents by inverting the flask at least ten times. If this is not done properly the solution will be progressively more concentrated towards the bottom of the flask.

Limits to the dissolving of solids

Saturated solutions

There is a definite limit to the amount of solute that will dissolve in a certain quantity of solvent at a particular temperature. For instance, if we add a little sodium chloride to water at room temperature, it will dissolve completely. If we add a little more, this also will dissolve. Eventually, however, there comes a point at which some sodium chloride crystals remain undissolved at the bottom of the vessel, and the solution is then said to be *saturated* at room temperature. The amount of undissolved sodium chloride has no effect on the concentration of the saturated solution. It does not matter whether there is just one crystal of undissolved solid or several hundred; at any given temperature the composition of a saturated solution is fixed. A saturated solution can therefore be described as one which, at a given temperature, will dissolve no more solute.

A *definition* of the term 'saturated solution' rests on the observation that the mass of solid in contact with a saturated solution remains constant. This is despite the fact that, over a long period of time, a crystal of undissolved solid gradually changes its shape. A change of shape suggests that the crystal may be dissolving away, at least in certain regions. Experiments have shown that this is true — but the crystal also becomes built up, in certain places, through crystallisation (see p. 88) of the solution, and the rate at which solid is deposited is exactly equal to the rate at which it dissolves. Consequently, the mass of solid remains constant, and the solid is said to be in *equilibrium*, *i.e.* a state of balance, with the solution. **A saturated solution can therefore be defined as one which, at a given temperature, is in equilibrium with undissolved solid.**

Solubility

A measure of the extent to which a solute will dissolve in a solvent at a particular temperature is provided by its *solubility*, which is the concentration of a saturated solution expressed on a mass/mass basis as grams of solute per 100 grams of *solvent* (not solution).

The term 'solubility' is therefore defined as **the number of grams of solute which will dissolve in 100 grams of the solvent at the temperature concerned so as to form a saturated solution.**

Solubility curves

We can prepare a saturated solution of, say, potassium nitrate in water at room temperature by putting crystals of potassium nitrate in a test tube, adding water and stirring the solution for about ten minutes until no more potassium nitrate dissolves. Crystals of undissolved solid must be present to ensure that the solution is saturated.

If we now heat the test tube with a bunsen burner the crystals of potassium nitrate are seen to dissolve, although a saturated solution can again be obtained by adding more potassium nitrate. Clearly, the solubility of potassium nitrate in water increases with temperature.

For any given solute and solvent, the variation of solubility with temperature is shown by a graph of solubility against temperature, known as a *solubility curve.*

The construction of a solubility curve for a salt in water lends itself to class participation, for each student (or pair of students) can determine the solubility of the salt at a given temperature between 20 °C and 95 °C (293–368 K) and then contribute the result towards the plotting of a curve.

The instructions below relate to the solubility curve of potassium nitrate, and should be followed by each student at his or her given temperature.

1. Collect the apparatus shown in Fig. 7.4, and pour 10 cm^3 of water into the test tube. The tube should be no more than half full; if it is too full, select a larger one. If we assume that the density of water at room temperature is 1 g cm^{-3}, then the mass of water introduced is 10 grams.

2. Place the test tube in the beaker of water, heat the beaker on a tripod and gauze, and hold it as steady as possible at a temperature two or three degrees below that for which the solubility is required.

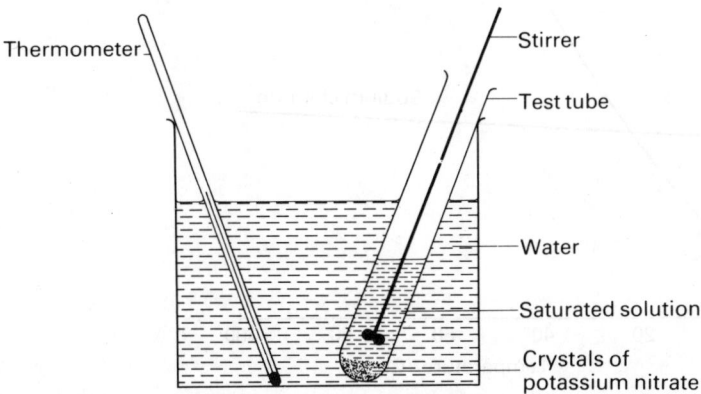

Fig. 7.4 Apparatus for the determination of the solubility of a salt in water

3. Add potassium nitrate crystals from a previously weighed stock until undissolved solid remains at the bottom of the tube. Reweigh the stock and hence find the mass of potassium nitrate that has been added.
4. Slowly heat the apparatus, with proper stirring of the solution, until the residual solid just dissolves. Note the temperature at this point.
5. As a check, allow the solution to cool, again with adequate stirring, until a cloudiness appears which indicates that the solution has begun to crystallise. Note the temperature, and take an average of this reading and the previous one.
6. Multiply the mass of potassium nitrate taken by ten to obtain the solubility of potassium nitrate, in grams per 100 grams of water, at the determined temperature.
7. Plot all the class results on a sheet of graph paper to obtain a solubility curve as shown in Fig. 7.5.

For many salts the variation of solubility with temperature is not as great as it is for potassium nitrate. The curve for sodium chloride, for instance, rises only gradually (see Fig. 7.5), and for certain compounds, such as calcium sulphate and calcium hydroxide, the solubility is actually lower in the hot than it is in the cold.

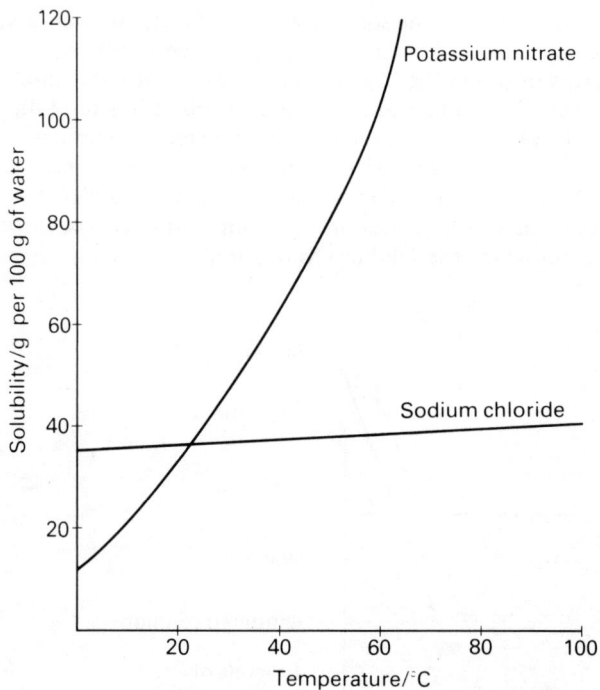

Fig. 7.5 Solubility curves for potassium nitrate and sodium chloride in water

Despite the fact that sodium chloride is only slightly more soluble in the hot than in the cold, it dissolves relatively quickly in hot water. It is important not to confuse *extent* of dissolving (*i.e.* solubility) with *rate* of dissolving.

Separation of solutions

Evaporation

This is the simplest means of recovering a dissolved solid from a solution. It relies on the fact that on heating the liquid is driven off while the solid remains behind. The only equipment that is required is an evaporating basin, a bunsen burner, a tripod and a gauze. The apparatus is arranged as shown in Fig. 7.6, and the solution is gently boiled to dryness.

Notes on evaporation

1. To avoid boiling over, the basin must not be overfilled.
2. Gentle heating is essential to avoid 'spitting' and consequent loss of solid. Some chemists minimise spitting by placing a clock glass over the basin, but this is not recommended because the glass tends to crack. It also reduces the rate of evaporation.
3. The bunsen burner must be turned out just before dryness is reached. If this is not done the temperature of the solid rises sharply, as a result of which it may spit violently or even decompose. Any liquid remaining on the hot solid evaporates as it cools down.
4. Evaporation, as described here, is an acceptable technique only for aqueous solutions, *i.e.* solutions in water. It must never be performed on solutions in alcohol, ether, etc., because the vapours given off by these liquids as they evaporate are both flammable and poisonous. However, such solutions can be evaporated safely on an electrically heated water bath placed in a fume cupboard. The absence of a naked flame lessens the risk of fire, while the extractor fan in the fume cupboard prevents pollution of the laboratory atmosphere and also keeps flammable vapours away from hot surfaces which might ignite them.

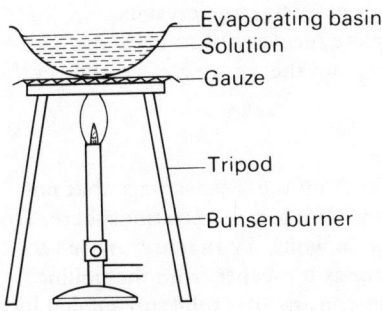

Fig. 7.6 The evaporation of a solution

Crystallisation

The technique of evaporation is open to two serious criticisms.

i.   Any dissolved impurities separate from the solution towards the end of the evaporation and contaminate the main solid.

ii. The inevitable rise in temperature at the end of the evaporation often causes some decomposition of the solid.

Both objections can be overcome by the adoption of a modified procedure known as *crystallisation*, whereby the solution is heated not to dryness, but only for as long as is necessary to give a concentrated solution. Judging this point is a skill acquired with practice, but basically involves looking for a ring of crystals around the side of the evaporating basin, and especially the formation of a slight 'skin' of solid on the surface. As the concentrated solution cools down to room temperature it *crystallises, i.e.* crystals of solid can be seen to 'grow' in the solution. Soluble impurities, as a general rule, remain in the *mother liquor, i.e.* the remaining solution. When crystallisation is complete, the solid is filtered off and washed, as it lies in the funnel, with a small amount of the pure liquid. This removes traces of the mother liquor. Finally the crystals are dried, first by pressing them between dry filter papers, and then by placing them on a watch glass or clock glass in a warm place.

Notes on crystallisation

1.   If crystals do not form well on cooling, it is likely that the solution is insufficiently concentrated. It should be reheated for a time, and then cooled again.

2.   If even a concentrated solution is reluctant to crystallise, it should be poured into a beaker standing in a freezing mixture of ice and water or, in desperate cases, ice and salt. Scratching the bottom of the beaker with a glass rod sometimes helps to start crystallisation, and another remedy, which is almost always successful, is to add a single crystal of the pure solid that is required. This acts as a nucleus on which other crystals can grow.

3.   The size of the crystals depends on the rate of crystallisation. Rapid cooling of a hot, concentrated solution causes the formation of a large number of small crystals, while the slow evaporation of a solution at room temperature leads to the growth of a few large crystals.

4.   Crystallisation can never give a complete recovery of solid. Some is bound to be lost in the mother liquor, and the washing of crystals in the funnel leads to a further loss.

Distillation

Both evaporation and crystallisation suffer from the disadvantage that one constituent of the solution, namely the liquid, is lost to the atmosphere. This is avoided by the technique of *distillation,* in which a *condenser* is used to condense (see p. 12) and collect the vapour as it escapes from the boiling solution. In its simplest form, a condenser consists of a tube surrounded by a cold water jacket (see Fig. 7.7).

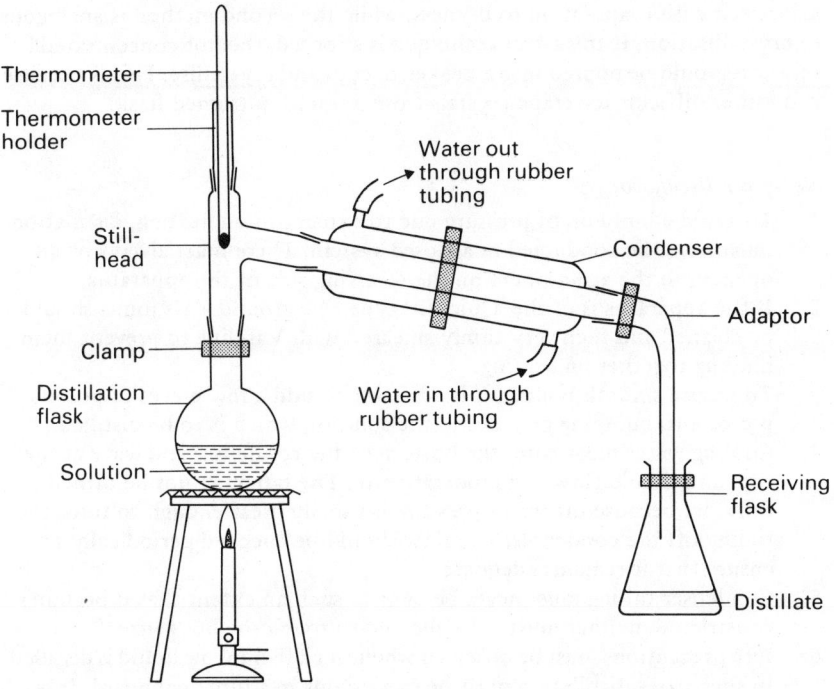

Fig. 7.7 Distillation apparatus (Shaded areas represent clamps)

The solution to be distilled is poured into a round bottomed distillation flask, so that the flask is no more than two-thirds full. The flask is clamped to a retort stand in such a way that it rests on a tripod and gauze, and a still-head is fitted to the top. If it is necessary to record the temperature of the vapour, a thermometer is placed in the still-head so that its bulb lies just below the level of the side arm. A condenser, to which rubber tubing has previously been affixed, is then provisionally clamped in roughly the right position, and carefully aligned with the side arm of the still-head. Only when the alignment is correct is the condenser finally clamped in position. Afterwards, an adaptor is fitted to the end of the condenser, and a receiving flask or beaker is finally placed as shown to collect the *distillate, i.e.* the pure liquid that distils over. It is important that the distillation apparatus is assembled in this order, for it is difficult to adjust the position of the distillation flask once the condenser has been fitted.

After the apparatus has been checked to ensure that all the joints are a good fit, the bunsen burner is lit and the solvent is distilled off at a moderate rate. Overheating must be avoided because it causes the solution to boil over and can lead, in certain cases, to thermal decomposition of the solute. The solution can either be distilled almost to dryness, or the bunsen burner can be turned out when the solution has become concentrated. The first method resembles evaporation (see above), and suffers from the usual drawbacks associated with evaporation to dryness, while the second method is analogous to crystallisation. If this latter technique is adopted, the hot concentrated solution should be poured into a beaker to cool and crystallise. This is because it is rather difficult to scrape crystals from a round bottomed flask.

Notes on distillation

1. To avoid a build-up of pressure due to expansion on heating, distillation must never be conducted in a closed system. There must always be an opening to the atmosphere on the receiving side of the apparatus.
2. If the apparatus is of the 'Quickfit' type, the ground glass joints should be cleaned and then very thinly smeared with Vaseline to prevent them binding together on heating.
3. To ensure smooth boiling, it is advisable to add a few pieces of porous pot or anti-bumping granules to the solution which is to be distilled.
4. Cooling water must enter the bottom of the condenser and leave at the top, and should flow at a moderate rate. The tap must not be turned fully on, because the water pressure is usually great enough to force the tubing off the condenser. The flow should be checked periodically to ensure that it remains adequate.
5. Condenser tubing must never be bent to such an extent that it becomes constricted; neither must it be allowed to touch the hot gauze.
6. Fire precautions must be observed whenever a flammable liquid is distilled. In such cases distillation must be carried out in a fume cupboard. It is advisable to use a hot water bath or an electric heating mantle as a source of heat, and boiling to dryness should be avoided. The receiver should be a flask, not a beaker, to minimise evaporation of the flammable distillate.

Fractional distillation

Distillation can be adapted for the separation of solutions of liquids in liquids. The technique, known as *fractional distillation,* entails the introduction of a vertical column called a *fractionating column* between the distillation flask and the still-head. The method is based on the principle that when a mixture of liquids is boiled, the vapour that escapes is nearly always richer in the most volatile constituent, *i.e.* the one which evaporates the most easily. For example, if a 50:50 mixture of petrol and oil (which, for the sake of argument, we can assume to be pure substances) is boiled, the vapour is mainly petrol. Condensation of this vapour gives a distillate that is mainly petrol. On repeated boiling and condensing in the fractionating column pure petrol is collected in the receiver, while pure oil remains behind in the distillation flask.

Fractional distillation can also be used for the separation of gases, such as the nitrogen and oxygen of the atmosphere. The mixture of gases to be separated is first liquefied by cooling and compression, and the liquid mixture is then fractionally distilled. In the case of liquid air, nitrogen boils off first, at $-195.8\ ^{\circ}C$ (77.4 K), followed by oxygen at $-183.0\ ^{\circ}C$ (90.2 K).

Summary

At the conclusion of this chapter, you should be able to:
1. recognise that when a solid dissolves in a liquid the molecules or ions of the solid become separated from one another,
2. define 'dissociation' as the separation of ions in an electrovalent compound,
3. define a 'solution' as a homogeneous mixture of a solute and a solvent,
4. explain the term 'suspension',
5. recognise that the concentration of a solution can be expressed in several ways, including mass/mass, mass/volume and molar mass/volume,
6. prepare a standard solution,
7. recognise that there is a limit to the mass of solute which can dissolve in a given mass of solvent,
8. explain the term 'saturated solution',
9. define the term 'solubility' in terms of mass of solute and mass of solvent,
10. recognise that solubility depends on temperature,
11. prepare a saturated solution,
12. define and use 'solubility curves',
13. separate solutions by crystallisation and distillation,
14. recognise that the component gases of the atmosphere can be separated by fractional distillation.

Questions

1. Which of the following statements are true and which are false?
   (a) A solution consists of a homogeneous mixture of a solute and a solvent.
   (b) For a solution of sugar in water, the sugar is the solute and the water is the solvent.
   (c) A solvent must always be a liquid at room temperature and atmospheric pressure.
   (d) The formation of a solution is an example of chemical change.
   (e) Ice dissolves at 0 °C (273 K) to give water.
   (f) Salt melts in water to give a solution commonly called brine.
   (g) Suspensions, unlike solutions, can be separated by filtration.
   (h) A molar solution is prepared by dissolving one mole of solute in 1 dm^3 of water.
   (i) The concentration of a saturated solution does not vary with temperature.
   (j) In the preparation of a standard solution, the graduated flask should be swilled out before use with distilled water.

In questions 2–11 the following relative atomic masses are required:
$H = 1, C = 12, N = 14, O = 16, Na = 23, S = 32, Cl = 35.5, K = 39, Cu = 63.5, Ag = 108, I = 127$.

In questions 2–4, calculate in each case how much solute must be weighed out in order to prepare the stated solution.

2. 250 cm^3 of sodium hydroxide solution of concentration 1 mol dm^{-3} (1 M).
3. 5 dm^3 of sodium chloride solution of concentration 4 mol dm^{-3} (4 M).
4. 100 cm^3 of sodium carbonate solution of concentration 0.05 mol dm^{-3} (0.05 M).

In questions 5–8, express the concentration of each solution in mol dm^{-3} (molarity).

5. Hydrochloric acid of mass concentration 30 g dm^{-3}.
6. Potassium iodide solution of mass concentration 100 g dm^{-3}.
7. Copper(II) sulphate solution containing 4 g of copper(II) sulphate per 250 cm^3 of solution.
8. Silver nitrate solution containing 1 g of silver nitrate per 100 cm^3 of solution.

In questions 9–11, calculate the mass concentration (*i.e.* the concentration in g dm^{-3}) of each solution.

9. Sulphuric acid of concentration 0.05 mol dm^{-3} (0.05 M).
10. Potassium hydroxide solution of concentration 0.1 mol dm^{-3} (0.1 M).
11. Potassium chloride solution of concentration 4 mol dm^{-3} (4 M).

In questions 12–15, select the most appropriate answer, labelled A, B, C or D.

12. The term 'dissociation' means the :
   A   formation of ions,
   B   separation of ions,
   C   joining together of ions,
   D   breakdown of molecules into atoms.
13. Which one of the following separation techniques is the most suitable for obtaining sea salt (impure sodium chloride) from sea water?
   A   Decantation.
   B   Filtration.
   C   Sublimation.
   D   Evaporation.
14. Which of the following techniques is the most suitable for obtaining relatively pure sodium chloride from sea water?
   A   Filtration.
   B   Sublimation.
   C   Evaporation.
   D   Crystallisation.
15. Which of the following techniques is the most suitable for obtaining pure water from sea water?
   A   Filtration.
   B   Distillation.
   C   Evaporation.
   D   Crystallisation.

For each of the mixtures shown in questions 16–18, select a separation technique from the list below which would give both constituents of the mixture in a relatively pure state.

A Filtration   C Fractional distillation
B Evaporation   D Sublimation

16. A suspension of chalk dust in water.
17. A solid mixture of salt and iodine.
18. A liquid mixture of water and ethanol (ethyl alcohol).
19. Use the following data to plot a graph of the solubility of potassium nitrate against temperature.

Temperature/$^\circ$C	20	30	40	50	60	70	80	90
Solubility/g per 100 g of water	31	45	62	84	110	139	171	215

Read off from the graph the solubility of potassium nitrate at (a) 35 $^\circ$C, (b) 67 $^\circ$C.

Chapter 8

Crystal structure

Many solid substances, *e.g.* common salt, sand, graphite, diamond and metals, are said to be *crystalline*, despite the fact that they may not have the appearance that we normally expect of crystals. The term 'crystalline' is used to denote the fact that in the solid state there is a regular and symmetrical arrangement of atoms, molecules or ions which extends throughout the whole of the crystal. The regular arrangement of atoms, molecules or ions is referred to as a *crystal lattice* or simply as a *lattice*.

A single crystal of a substance has a definite, symmetrical shape, *e.g.* a cube for sodium chloride. However, many substances, *e.g.* metals, consist of a very large number of crystals assembled together, and in such cases the material can assume any shape.

We can divide crystalline substances into two broad categories, namely *molecular crystals* and those possessing *giant lattices*.

Molecular crystals are formed by substances which are composed of simple molecules, while giant lattice structures are formed by three types of substances, as follows.

i.   Those which consist of giant molecules (see p. 46). The crystals formed in such cases are called *atomic crystals*.
ii. Ionic (*i.e.* electrovalent) compounds, which give *ionic crystals*.
iii. Metals, which give *metallic crystals*.

Molecular crystals

The lattices of most molecular crystals comprise a regular arrangement of molecules, as illustrated by part of the iodine structure in Fig. 8.1(a). The

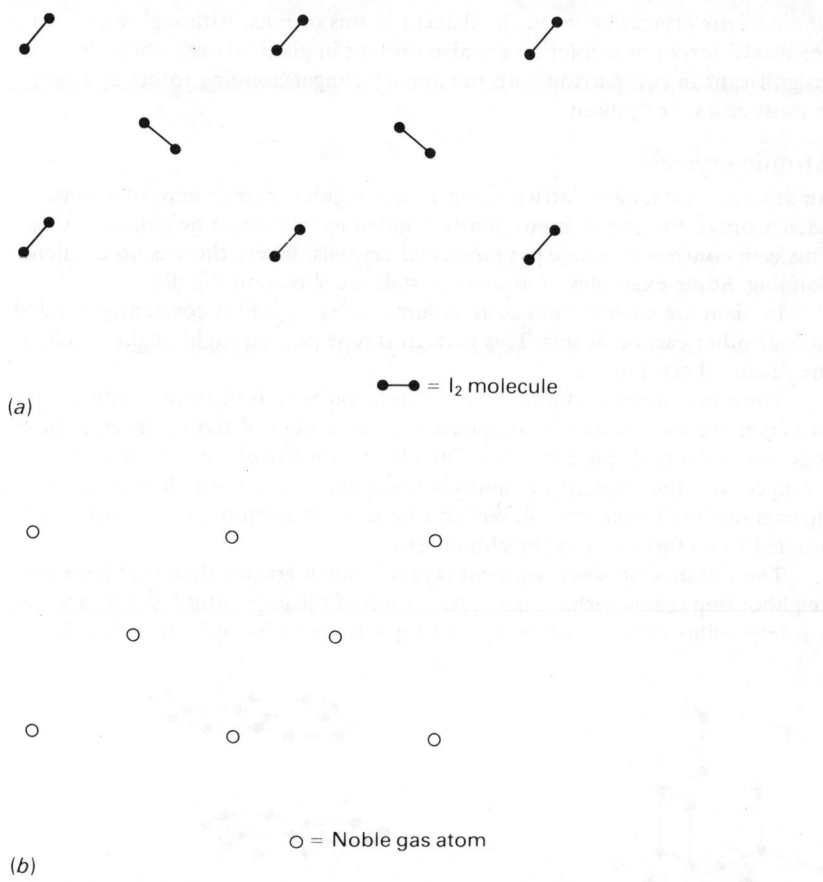

(a)

●—● = I₂ molecule

○ = Noble gas atom

(b)

Fig. 8.1 The arrangement of (*a*) molecules in a crystal of iodine, and (*b*) atoms in a solidified noble gas

solidified noble gases (see Fig. 8.1(b)) form similar crystals, in which there is a regular arrangement of non-covalently bonded atoms. All such substances have many properties in common and are referred to as 'molecular crystals' regardless of whether they are built up of molecules or atoms.

The molecules (or noble gas atoms) are held closely together in a molecular crystal by weak forces of attraction, called *van der Waals' forces*. These weak *inter*molecular forces, *i.e.* forces between neighbouring molecules, contrast with the strong *intra*molecular forces, *i.e.* forces which exist between neighbouring atoms within a particular molecule. The latter, as we have seen, are termed 'covalent bonds'.

Giant lattices

In every giant lattice structure strong bonding forces exist throughout the whole of the crystal between all adjacent atoms or ions. Although weak van der Waals' forces of attraction are also present in giant lattices, they are insignificant in comparison with the much stronger bonding forces and can, in most cases, be ignored.

Atomic crystals

An atomic crystal has a lattice comprising a regular arrangement of atoms. Each atom in the lattice is covalently bonded to its nearest neighbour atoms. This is in contrast to noble gas molecular crystals, where there is no covalent bonding. Some examples of atomic crystals are shown in Fig. 8.2.

In diamond each carbon atom is surrounded by and is covalently bonded to four other carbon atoms. This pattern is repeated throughout the whole of the diamond crystal.

The structure of graphite, which also involves carbon atoms, is described as a *layer lattice*, because it comprises a large number of flat layers stacked together like a multiple sandwich. Each layer consists of carbon atoms so arranged that they describe a multiple hexagonal pattern which extends throughout the whole crystal. Within a layer, each carbon atom is covalently bonded to its three nearest neighbour atoms.

The distance between adjacent layers is much greater than that between neighbouring atoms within a layer. As a result of the large inter-layer separation, it is impossible for the carbon atoms of one layer to form covalent bonds

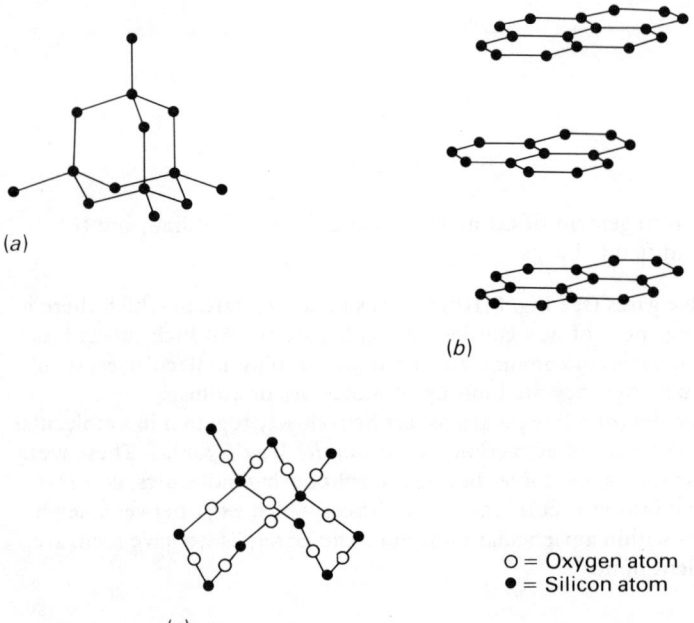

(a)

(b)

(c)

O = Oxygen atom
● = Silicon atom

Fig. 8.2 The arrangement of atoms in (a) diamond, (b) graphite and (c) silica

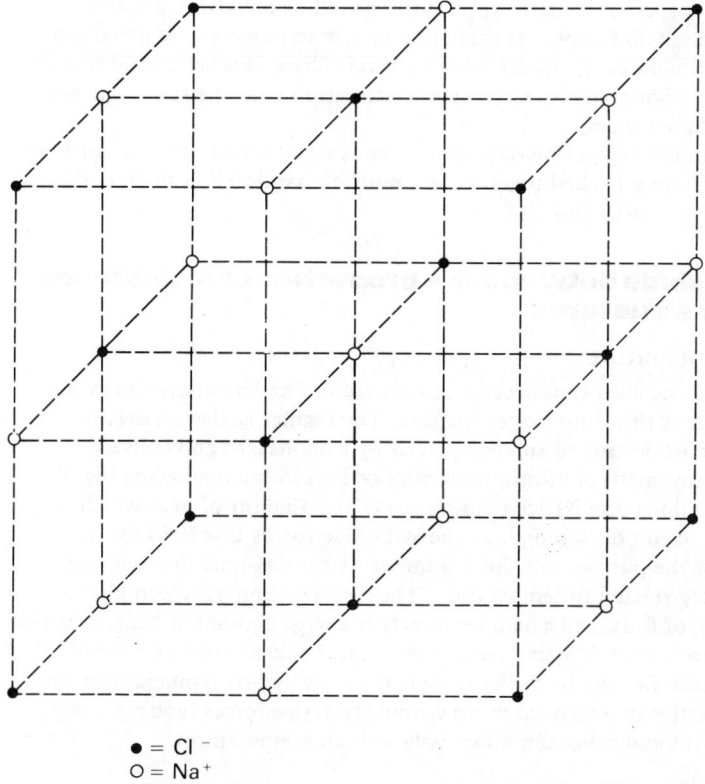

with those of another. The only attraction between adjacent layers is provided by van der Waals' forces. A crystal of graphite, however, is not classified as a molecular crystal since each layer in graphite is very much larger than any molecule in a molecular crystal.

In silica (otherwise known as silicon dioxide), a substance which occurs naturally as sand, each atom of silicon is surrounded by and covalently bonded to four atoms of oxygen. Each oxygen atom in turn is surrounded by and bonded to two silicon atoms. This arrangement is possible because, as the formula, SiO_2, indicates, there are twice as many oxygen atoms as silicon atoms.

To summarise, the principal binding forces which hold atomic crystals together are the covalent bonds between the atoms.

Ionic crystals

The lattice of an ionic crystal consists of a regular arrangement of ions. Ions of like charge repel one another, while those of opposite charge are attracted together. The ions of an ionic crystal are packed together in such a way as to

● = Cl⁻
O = Na⁺

Fig. 8.3 The arrangement of ions in a crystal of sodium chloride

minimise repulsion and maximise attraction. The nearest neighbours to any ion within an ionic crystal are therefore ions of opposite charge; not ions of like charge.

There are various possible arrangements of ions in the crystal lattice. For any ionic compound the arrangement is determined by two things:
i. the relative sizes of the positively charged and negatively charged ions,
ii. the relative charges on the ions, and hence the ionic ratio.

A part of the sodium chloride lattice is shown in Fig. 8.3. In this figure **the dashed lines do not represent bonds**; they simply show that the ions are packed together in such a way as to describe a cube. Notice that each sodium ion is surrounded by six near neighbour chloride ions, and that each chloride ion is similarly surrounded by six sodium ions. This arrangement extends throughout the whole of the crystal.

To summarise, the forces which bind ionic crystals together are the ionic bonds, *i.e.* the electrostatic forces of attraction between oppositely charged ions.

Metallic crystals

As the name suggests, this term applies to crystalline metals. In a metallic crystal, as we saw in Chapter 4, there is a regular arrangement of metal ions extending throughout the whole of the crystal. These ions are bound together by the metallic bond, *i.e.* by electrostatic attraction with the sea of valency electrons between them.

One common type of metallic crystal consists of layers of metal ions (see Fig. 8.4(a)), closely packed together as a multiple sandwich in the metallic crystal (see Fig. 8.4(b) and (c)).

Relationships between the properties of substances and their structures

Melting temperature

In any crystal the atoms, molecules or ions are in fixed positions, but when the crystal melts this is no longer the case. The regular lattice arrangement is broken down or destroyed and is replaced by a haphazard and constantly changing arrangement of atoms, molecules or ions in the liquid (see Fig. 8.5).

Breaking down the lattice requires heat. The amount of heat which is needed depends on the strength of the attractive forces that hold the lattice together. For the purposes of this argument we may assume that amount of heat is directly related to temperature. Thus, a low temperature implies a small amount of heat, and a high temperature a large amount of heat. A lattice in which there are weak attractive forces can be broken down by a small amount of heat, *i.e.* can be broken down at relatively low temperatures. In contrast, a lattice in which there are strong attractive forces requires a large amount of heat and is broken down only at high temperatures.

Molecular crystals

To melt a molecular crystal sufficient heat is needed to overcome the inter-

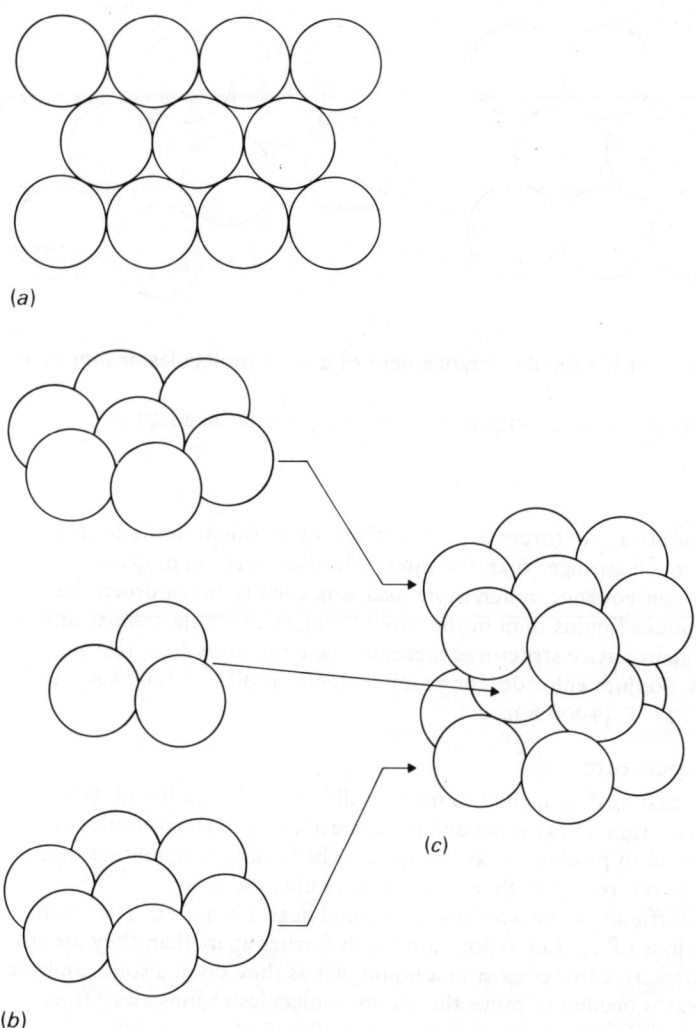

Fig. 8.4 The arrangement of ions in a crystalline metal
(*a*) Close packed metal ions forming a layer
(*b*) A side view of parts of three layers of the type shown in (*a*)
(*c*) The layers from (*b*) packed closely together in the metallic crystal

molecular forces only, so that the molecules (or noble gas atoms) can move away from their positions in the lattice to produce the random arrangement of the liquid. The intermolecular forces are weak and as a result are overcome by a relatively small amount of heat. Consequently, molecular crystals often possess low melting temperatures. Water, for example, melts at 0 °C (273 K), and argon at −189 °C (84 K).

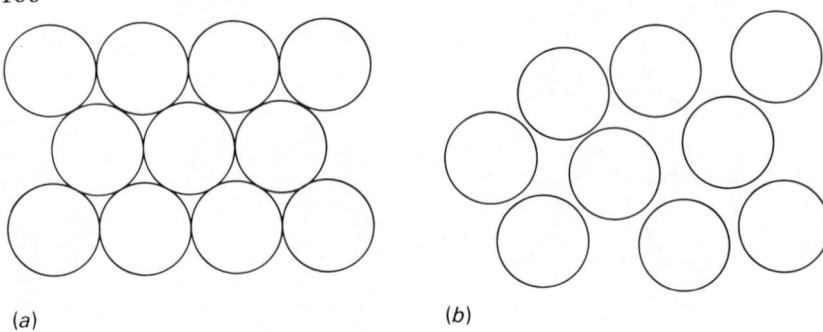

(a)   (b)

Fig. 8.5 (a) Part of the regular arrangement of atoms, molecules or ions in a crystal
(b) The rapidly changing arrangement of similar particles in a liquid

Giant lattices

The binding or attractive forces in giant lattices, *i.e.* covalent, ionic or metallic bonding, are much stronger than the intermolecular forces in molecular crystals. As a consequence, much more heat is needed to break down the lattices to produce liquids than in the case of molecular crystals. Most substances with giant lattice structures therefore have relatively high melting temperatures. Sodium chloride, for example, melts at 801 °C (1074 K), and graphite at 3727 °C (4000 K).

Boiling temperature

We have seen that heat is needed to melt a solid to produce a liquid. When a liquid boils, additional heat is needed to separate the atoms, molecules or ions in the liquid to produce a gas or vapour. The boiling temperature of a substance is therefore higher than its melting temperature.

It is not difficult to see why this additional heat is required. The atoms, molecules or ions of a gas or vapour are much further apart than they are in the liquid. Attractive forces exist in a liquid just as they do in a solid, and the additional heat is needed to move the atoms, molecules or ions away from one another against these attractive forces (see Fig. 8.6).

The attractive forces in a liquid are similar in nature to those in the crystal from which it was formed. Thus, molecular liquids, *i.e.* those obtained by melting molecular crystals, have much lower boiling temperatures than liquids which are produced by melting substances with giant lattice structures.

Electrical conductivity

Substances through which electricity can flow are referred to as *electrical conductors* or simply as *conductors*, while those through which electricity cannot flow are known as *non-conductors*. The passage of electricity is associated with the movement of charged particles through the conductor. For a substance to be a conductor it must therefore contain charged particles which can move about freely. These charged particles may be electrons, in the case of a metal, or ions, in the case of a salt.

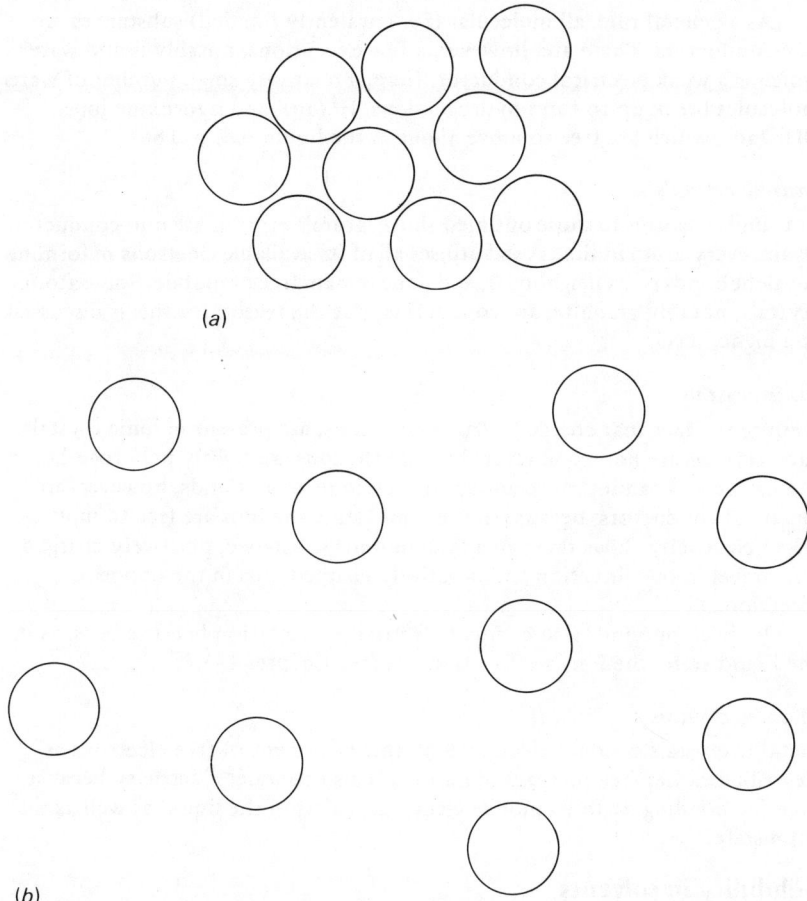

Fig. 8.6 (*a*) The random arrangement of atoms, molecules or ions in a liquid (*b*) Separated atoms, molecules or ions in a gas or vapour

If there are no charged particles present, or if charged particles do exist but are unable to move, the material is a non-conductor.

Molecular crystals

In all simple molecules every atom uses all its available electrons in forming covalent bonds, and there are no free (or mobile) electrons left over. In an iodine molecule, for example, each iodine atom is covalently bonded to another iodine atom, while in a molecule of tetrachloromethane each carbon atom is covalently bonded to four chlorine atoms:

$$I–I \qquad \begin{array}{c} Cl \\ | \\ Cl–C–Cl \\ | \\ Cl \end{array}$$

Consequently, these substances are non-conductors of electricity.

As a general rule, all molecular (*i.e.* covalently bonded) substances are non-conductors. There are, however, a few exceptions, notably liquid water, which is a weak electrical conductor. Here, a relatively small number of water molecules break up to form hydrogen ions, H^+(aq), and hydroxide ions, OH^-(aq), which are free to move about in the liquid (see p. 188).

Atomic crystals

For similar reasons to those outlined above, atomic crystals are non-conductors. Again, every atom in the crystal utilises all of its available electrons in forming covalent bonds to its neighbours, and none remain free or mobile. Some atomic crystals, notably graphite, are conductors, but the reason for this is discussed at a higher level.

Ionic crystals

Despite the fact that charged particles, *i.e.* ions, are present in ionic crystals, such crystals are non-conductors because the ions are tightly held together in the lattice and cannot move about. *Molten* ionic compounds, however, are electrical conductors, because in the liquid state the ions are free to move. When electricity flows through a molten ionic substance, positively charged ions travel in one direction and negatively charged ions in the opposite direction.

Ionic compounds also conduct electricity in solution because here, as in the liquid state, the ions are free to move (see Chapter 14).

Metallic crystals

Metallic crystals conduct electricity by the movement of free electrons as described earlier (see p. 37). Liquid metals also conduct electricity because metallic bonding, with its mobile electrons, exists in the liquid as well as the solid state.

Solubility in solvents

When a crystalline solid dissolves in a solvent, the lattice is broken down and a random arrangement of atoms, molecules or ions is produced in solution. The main factors governing solubility are discussed at a higher level, but some important points can be made here.

Solvents fall broadly into two categories, namely *polar* and *non-polar*. Water is said to be 'polar' because the oxygen atom in a water molecule carries an appreciable negative charge (δ^-), while the hydrogen atoms each carry positive charges (δ^+):

$$\overset{\delta^+}{H} \diagdown \overset{\delta^-}{\underset{O}{}} \diagup \overset{\delta^+}{H}$$

Tetrachloromethane, in common with many organic solvents, is said to be 'non-polar' because there is little or no charge on any of the atoms.

Solutes, also, can be divided into two groups: (i) those which consist of ions or polar molecules, and (ii) those which consist of non-polar molecules. It is a frequently observed rule that 'like dissolves like'. Polar solutes dissolve in polar solvents, non-polar solutes dissolve in non-polar solvents, but the solubility of polar solutes in non-polar solvents (or of non-polar solutes in polar solvents) is generally low.

Water, for example, but not tetrachloromethane, is a good solvent for many ionic crystals, and it is not difficult to see why this is so. The positive hydrogen atoms of the water molecules attract the negative ions in a crystal, while the negative oxygen atoms attract the positive ions. A non-polar tetrachloromethane molecule, however, does not attract the ions in a crystal in any way.

If the molecules that make up a molecular crystal are non-polar, *e.g.* iodine, I_2, they resemble tetrachloromethane molecules rather than water molecules. Hence iodine is readily soluble in tetrachloromethane but is only sparingly (*i.e.* slightly) soluble in water. However, if a molecular crystal is composed of polar molecules, *e.g.* sugar, water rather than tetrachloromethane will be a suitable solvent for the crystal.

Atomic crystals do not, as a general rule, dissolve in any solvents, polar or non-polar. The reason for this is that the solvent molecules are unable to break the strong covalent bonds between atoms in the crystal. There are two forms of phosphorus which provide a good illustration of this. White phosphorus, which forms molecular crystals, dissolves in non-polar solvents, but red phosphorus, which forms atomic crystals, is insoluble in all solvents.

Metals, in general, do not dissolve in solvents unless chemical reactions take place.

Summary

At the conclusion of this chapter, you should be able to:
1. define a 'lattice' as a regular arrangement of atoms, molecules or ions,
2. state that solid substances may be divided broadly into those forming simple molecular crystals, and those possessing giant lattices,
3. state examples of molecular crystals, *e.g.* iodine,
4. state that molecules are weakly attracted to one another by van der Waals' forces,
5. classify solids having giant lattice structures as
   (*a*) atomic crystals, *e.g.* diamond or silica,
   (*b*) ionic crystals, *e.g.* sodium chloride,
   (*c*) metallic crystals,
6. recognise structures such as sodium chloride, diamond or graphite from diagrams or models,
7. describe how ions are held together in an ionic crystal,
8. recognise that low melting temperatures and boiling temperatures are usually associated with simple molecular structures, and high melting temperatures and boiling temperatures with giant lattice structures,
9. state the difference in electrical conductivity of molecular crystals,

atomic crystals, ionic crystals (solid and molten) and metallic crystals,
10. explain these differences in terms of mobile electrons or ions,
11. relate the non-conducting properties of atomic crystals to the presence of covalent bonds,
12. state that ionic substances such as sodium chloride are soluble in water but insoluble in many non-aqueous solvents,
13. state that simple molecular substances such as iodine are often sparingly soluble in water but freely soluble in organic solvents.

Questions

1. Classify each of the following substances in the solid state as a molecular crystal, atomic crystal, ionic crystal or metallic crystal.

   (a) Mercury   (f) Sodium
   (b) Neon   (g) White phosphorus
   (c) Silica   (h) Oxygen
   (d) Bromine   (i) Potassium chloride
   (e) Red phosphorus   (j) Graphite

For each of the substances named in questions 2–6, select *all* the appropriate properties from the list below.
A   Relatively low melting temperature
B   Relatively high melting temperature
C   Soluble in water
D   Insoluble in water
E   Electrical conductor in the molten and solid states
F   Electrical conductor in the molten but not the solid state
G   Electrical non-conductor under all conditions
2. Argon
3. Graphite
4. Sodium chloride
5. Iron
6. Iodine

In questions 7–13, select the most appropriate answer, labelled A, B, C or D.
7. In an atomic crystal there is a regular arrangement of:
   A   atoms which are not bonded together,
   B   atoms which are bonded together throughout the crystal,
   C   ions,
   D   metal atoms.
8. In a crystal of sodium chloride, each:
   A   ion has a charge of +1,
   B   sodium ion is surrounded by six chloride ions,
   C   chloride ion is surrounded by four sodium ions,
   D   sodium atom is covalently bonded to a chlorine atom.

9. Which one of the following has a layer lattice?
   A Diamond.
   B Silica.
   C Graphite.
   D Argon.

10. Ionic crystals have high melting temperatures because:
    A the attractive forces between ions are strong,
    B the molecules are stable,
    C of the strong covalent bonds that exist within them,
    D of the weak van der Waals' forces within them.

11. Which one of the following conducts electricity in the molten state but not in the solid state?
    A Sodium chloride.
    B Graphite.
    C Diamond.
    D Zinc.

12. Water is a weak electrical conductor because:
    A there are free electrons present,
    B the water molecules possess electrical charges,
    C a few water molecules break up into ions,
    D all the water molecules are split into ions.

13. Which one of the following substances is appreciably soluble in water?
    A Silica.
    B Red phosphorus.
    C Graphite.
    D Sodium chloride.

Chapter 9

Chemical reactions

Many chemical changes are not as simple as they appear to be at first sight. This is because they consist of a number of separate changes, which can occur at the same time as one another or in rapid succession. Each of these individual changes is known as a *chemical reaction*.

The burning of methane (natural gas) is a particularly simple chemical change, for it consists of one chemical reaction only. In this reaction, methane combines with the oxygen of the air to give carbon dioxide and water. However, if methane is burnt in a limited supply of air, there is still a chemical change, but this time it involves *two* chemical reactions. In one, methane combines with oxygen to give carbon dioxide and water as before. In the other, methane combines with oxygen to give carbon *monoxide* and water. Carbon monoxide is a very poisonous gas; that is why it is so important to ensure that there is always a plentiful supply of air to gas appliances.

Most chemical changes in nature are extremely complicated because they involve numerous chemical reactions. On an elementary chemistry course we confine ourselves exclusively to studying chemical changes which take place by one reaction only.

The general principles of chemical reactions are discussed in this chapter. Some particular reactions are described in Chapters 10–13, while Chapter 14 (Electrochemistry) is concerned with chemical reactions which are brought about by means of electricity.

Features of chemical reactions

Law of Conservation of Mass

The substances that take part in a chemical reaction are called *reactants*, while the new substances that are formed are called *products*. It is a basic law of chemistry that, when a chemical reaction occurs, the total mass of the products is equal to the total mass of the reactants.

This principle is known as the *Law of Conservation of Mass*. It was first proposed in 1756 by Mikhail Lomonósov, and confirmed a few years later by Antoine Lavoisier. In 1893 Hans Landolt conducted a most accurate series of experiments and again demonstrated the truth of this law. To exclude the possibility of air being absorbed by the reactants, or gases being given off to the atmosphere, Landolt carried out his experiments in a closed glass vessel now known as a 'Landolt flask' (see Fig. 9.1).

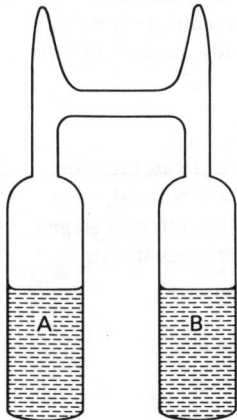

Fig. 9.1 A Landolt flask

In one experiment a solution of sodium chloride was introduced into bulb A, and silver nitrate solution into B, after which the necks of both bulbs were sealed by a flame. The flask was then weighed on an accurate balance and inverted. The two solutions intermingled so that a chemical reaction occurred to give a white solid called silver chloride and a solution of sodium nitrate. The flask was finally reweighed, but no difference in mass could be detected.

There is no logical proof of this law. It is merely an observed fact that atoms are neither created nor destroyed during chemical changes.

Exothermic and endothermic reactions

Some chemical reactions, for example that between zinc and sulphuric acid, begin immediately the reactants come into contact with each other. Others, for example the burning of coke in oxygen, do not occur until heat or some other form of energy is supplied. (Anyone who has ever struggled to light a domestic fire will appreciate this.) **Once they have started**, however, all reactions are accompanied by a heat change, and can be divided into two groups depending on whether heat is given out or taken in. In *exothermic* reactions (Greek *exo* = 'outside', *thermos* = 'hot') heat is lost to the surroundings, while in *endothermic* reactions (Greek *endon* = 'within') heat is absorbed from the surroundings.

Both the reactions quoted above (*i.e.* the dissolving of zinc in sulphuric acid, and the combustion of coke) are exothermic. Granted, the burning of coke needs heat to get it going, but once the reaction has started heat is given out. Endothermic reactions, in contrast, require a continuous supply of heat. An example is the decomposition of limestone (calcium carbonate) in a lime kiln (see Fig. 13.1). A temperature of about 800 °C (1073 K) is needed, and if this high temperature is not maintained the reaction ceases.

The heat change that accompanies a chemical reaction is called an *enthalpy change*, from the Greek verb *enthalpein* which means 'to heat'. It has its origin in the making and breaking of chemical bonds. Essentially, there are two stages to any reaction:
i. the breaking of chemical bonds in the reactants,
ii. the formation of chemical bonds in the products.

Stage i does not occur by itself. Energy, in the form of heat, must be supplied. Stage ii, by contrast, occurs entirely of its own accord and, as in all such processes, energy is released. If the heat which is liberated in stage ii is greater than that which is required in stage i the reaction is exothermic; otherwise it is endothermic.

Chemical equations

Molecular equations

A chemical reaction can be represented by a *molecular equation* that shows, in their correct proportions, the various atoms, molecules or ions that take part, together with those that are formed. The reactants are written on the left-hand side of the equation and the products on the right. Thus, the equation:

$$Hg + Cl_2 = HgCl_2$$

indicates that one atom of mercury reacts with one molecule of chlorine to give one molecule of mercury(II) chloride. If we wish to emphasise that liquid mercury reacts with chlorine gas to form solid mercury(II) chloride, we write:

$$Hg(l) + Cl_2(g) = HgCl_2(s)$$

The following symbols are used in the writing of chemical equations.
i. + The plus sign is largely self-explanatory. When used on the left-hand side of an equation, it means 'combines with'.
ii. = This means 'yield', *i.e.* 'react to form'.
iii. (s), (l), (g) and (aq) are used when necessary to denote the physical states of the reactants and products. (s) stands for solid, (l) for liquid, (g) for gas and (aq) for substances in aqueous solution.

Every chemical equation must *balance*, *i.e.* for each element involved, the number of atoms on the left-hand side of the equation must be equal to that on the right. This ensures that the total mass is the same on both sides of the equation, in accordance with the Law of Conservation of Mass.

Some equations, like that above, balance immediately the formulae of the substances involved are written down, but others do not. Consider the reaction between hydrogen and chlorine to produce hydrogen chloride. If we merely write down the formulae of the substances the equation does not balance. We therefore cannot write it with an '=' sign but can use an arrow instead:

$$H_2 + Cl_2 \rightarrow HCl$$

There are two hydrogen and two chlorine atoms on the left-hand side, but only one of each on the right. To make the equation balance we therefore need two molecules of hydrogen chloride:

$$H_2 + Cl_2 = 2HCl$$

We now have equal numbers of hydrogen and chlorine atoms on both sides of the equation, and we are entitled to use the = sign.

All chemical equations can be balanced by trial and error. The general technique is to write down the formulae of the substances involved, and then multiply certain of them by low integers until atoms of all the elements balance. Consider, for example, the reaction between carbon and iron(III) oxide to give iron and carbon monoxide:

$$C + Fe_2O_3 \rightarrow Fe + CO$$

Iron and oxygen are out of balance. This can be remedied by multiplying Fe by 2 and CO by 3:

$$C + Fe_2O_3 \rightarrow 2Fe + 3CO$$

A check shows that we now have three carbon atoms on the right-hand side but only one on the left. The balance can readily be restored by multiplying C on the left by 3:

$$3C + Fe_2O_3 = 2Fe + 3CO$$

As a more complicated example, take the reaction of aluminium with hydrochloric acid to give aluminium chloride and hydrogen:

$$Al + HCl \rightarrow AlCl_3 + H_2$$

Neither hydrogen nor chlorine balances. There is more than one way of balancing the equation, but we could proceed as follows.

Multiply HCl by 2 to make hydrogen balance:

$$Al + 2HCl \rightarrow AlCl_3 + H_2$$

Multiply 2HCl by 3 and $AlCl_3$ by 2 to make chlorine balance:

$$Al + 6HCl \rightarrow 2AlCl_3 + H_2$$

Multiply Al by 2 to make aluminium balance:

$$2Al + 6HCl \rightarrow 2AlCl_3 + H_2$$

Multiply H_2 by 3 to make hydrogen balance:

$$2Al + 6HCl = 2AlCl_3 + 3H_2$$

Few equations are as difficult to balance as this. Very often a short cut can be taken by balancing covalent ions (see p. 52) rather than individual elements. For example, consider the reaction between chromium(III) sulphate and sodium hydroxide to give chromium(III) hydroxide and sodium sulphate:

$$Cr_2(SO_4)_3 + NaOH \rightarrow Cr(OH)_3 + Na_2SO_4$$

Multiply NaOH by 3 to make hydroxide ions balance, and multiply Na_2SO_4 by 3 to make sulphate ions balance:

$$Cr_2(SO_4)_3 + 3NaOH \rightarrow Cr(OH)_3 + 3Na_2SO_4$$

Multiply $Cr(OH)_3$ by 2 to make chromium balance, and 3NaOH by 2 to make sodium balance:

$$Cr_2(SO_4)_3 + 6NaOH = 2Cr(OH)_3 + 3Na_2SO_4$$

In general, we do not make a conscious effort to remember balanced equations. We memorise only the substances involved, and balance the equations as required.

Ionic equations

In chemical reactions which involve ionic substances, it is seldom that all the ions play a direct part. Those that do not are termed *spectator ions*. Consider, for example, the reaction between sodium chloride and silver nitrate. Reaction occurs in aqueous solution to give a white precipitate of silver chloride; at the same time sodium nitrate is formed and remains in solution. We can write a molecular equation for the reaction:

$$NaCl + AgNO_3 = AgCl + NaNO_3$$

Sodium chloride exists in solution as sodium ions, Na^+, and chloride ions, Cl^-. Silver nitrate provides silver ions, Ag^+, and nitrate ions, NO_3^-. When the solutions are mixed together silver ions join with chloride ions to give solid silver chloride, while the spectator ions, Na^+ and NO_3^-, remain in solution. We can ignore these spectator ions and write an *ionic equation* to show only those ions which participate directly in the reaction:

$$Ag^+ + Cl^- = AgCl$$

We can make the ionic equation more informative by using the notation introduced earlier:

$$Ag^+(aq) + Cl^-(aq) = AgCl(s)$$

Ionic equations can be written for any type of reaction in which ions are involved.

Stoicheiometry of reactions

A balanced chemical equation shows the ratio in which molecules (or atoms or ions, as the case may be) react together. For example, the equation:

$N_2 + 3H_2 = 2NH_3$

indicates that one molecule of nitrogen combines with three molecules of hydrogen to give two molecules of ammonia. The *molar* ratio for the reaction is exactly the same as the molecular ratio (see p. 70), *i.e.*

1 mole N_2 + 3 moles H_2 = 2 moles NH_3

The relative atomic masses of hydrogen and nitrogen are 1 and 14 respectively, so that the relative molecular masses of N_2, H_2 and NH_3 are 28, 2 and 17 respectively:

therefore 1 mole of N_2 has a mass of 28 g,
           3 moles of H_2 have a mass of 6 g,
and   2 moles of NH_3 have a mass of 34 g,
therefore 28 g N_2 + 6 g H_2 = 34 g NH_3

In accordance with the Law of Conservation of Mass, the total masses on both sides of the equation are equal.

The mole concept can be extended to include electrovalent compounds, *e.g.*

$AgNO_3 + KCl = AgCl + KNO_3$
1 mole   1 mole   1 mole 1 mole

The relative atomic masses of N, O, Cl, K and Ag are 14, 16, 35.5, 39 and 108 respectively, so that the relative molecular masses of $AgNO_3$, KCl, AgCl and KNO_3 are 170.0, 74.5, 143.5 and 101.0 respectively,

therefore 170.0 g $AgNO_3$ + 74.5 g KCl = 143.5 g AgCl + 101.0 g KNO_3

In this way we can calculate the mass of a substance that will react with a certain mass of another substance. We can also calculate the *theoretical yield* of each of the products, *i.e.* the mass that would be formed if the reaction went to completion with no losses of any kind. Such calculations have to be made before we can attempt any chemical preparation in the laboratory or in industry, for we must know what quantities of reactants to use in order to obtain the required amount of product.

Calculations on reacting masses are known as *stoicheiometric calculations*, the term *stoicheiometry* being defined as 'the branch of chemistry that deals with the proportions in which substances react' (Chambers's Dictionary). Molecular equations, because they provide the basis for these calculations, are referred to as *stoicheiometric equations*.

The basic procedure to be followed in performing stoicheiometric calculations is as follows.
i.   Write down the correctly balanced chemical equation. For example, to find the mass of calcium oxide formed on heating 50 g of calcium

carbonate, we begin by writing the equation:

$$CaCO_3 = CaO + CO_2$$

ii. Write down any given mass (or masses) in grams, and convert to an amount in moles by dividing by relative molecular mass.

      Here, we write down 50 g of calcium carbonate and then divide by 100, which is the relative molecular mass of this compound. This gives us 0.5 mol of calcium carbonate.

iii. Use the equation to calculate the required amount of substance in moles, and finally reconvert to grams by multiplying by relative molecular mass.

      In our case, the equation tells us that one mole of calcium carbonate, on decomposition, gives one mole of calcium oxide. Therefore, 0.5 mol of calcium carbonate gives 0.5 mol of calcium oxide. To convert to grams, we multiply by the relative molecular mass of calcium oxide, *i.e.* by 56. This gives a final answer of 28 g of calcium oxide.

To gain a fuller understanding of the method we shall now study some other examples.

Reactions involving one reactant only

No modification of the basic procedure is necessary.

Example

Calculate the masses of lead(II) oxide, nitrogen dioxide and oxygen that are formed by heating 165.5 g of lead(II) nitrate. (Relative atomic masses: $N = 14, O = 16, Pb = 207$)

Answer

i. The balanced equation is as follows:

$$2Pb(NO_3)_2 = 2PbO + 4NO_2 + O_2$$

ii. We have 165.5 g of lead(II) nitrate. Its relative molecular mass is $207 + 28 + 96 = 331$

      therefore we have $\dfrac{165.5}{331}$ = 0.5 mol of lead(II) nitrate

iii. From the equation,

                2 mol $Pb(NO_3)_2$ gives 2 mol PbO + 4 mol NO_2 + 1 mol O_2,

      therefore 0.5 mol $Pb(NO_3)_2$ gives 0.5 mol PbO + 1 mol NO_2 + 0.25 mol O_2,

      i.e. the product is 0.5 × 223 g PbO = 111.5 g PbO,

                           1 × 46 g NO_2 = 46.0 g NO_2,

      and   0.25 × 32 g O_2 = 8.0 g O_2.

      The total mass of the products is 165.5 g. This tallies with the original mass of lead(II) nitrate and provides a useful check on the calculation.

Reactions involving two reactants

If two (or more) reactants are mixed together in the molar ratio shown in the equation for the reaction, the calculation of the yield can successfully be based on either of the reactants. For example, suppose we mix together 31.5 g of nitric acid and 20.0 g of sodium hydroxide. How much sodium nitrate is formed?

$$HNO_3 + \quad NaOH = NaNO_3 + H_2O$$

31.5 g \qquad 20.0 g

$$= \quad \frac{31.5}{63} \text{ mol} \quad \frac{20.0}{40} \text{ mol}$$

$$= \quad 0.5 \text{ mol} \quad 0.5 \text{ mol}$$

The reactants are therefore present in an equimolar ratio as required by the equation. We could argue (A) that

1 mol HNO_3 gives 1 mol $NaNO_3$,

therefore 0.5 mol HNO_3 gives 0.5 mol $NaNO_3$,

$$= \quad 0.5 \times 85 = 42.5 \text{ g } NaNO_3$$

Alternatively, we could argue (B) that

1 mol $NaOH$ gives 1 mol $NaNO_3$,

therefore 0.5 mol $NaOH$ gives 0.5 mol $NaNO_3$,

$$= \quad 42.5 \text{ g } NaNO_3$$

But how much sodium nitrate is formed if we mix together, say, 20.0 g of sodium hydroxide and 131.5 g of nitric acid? Is it still 42.5 g by argument B, or can we adapt argument A as follows?

$$HNO_3 + NaOH = NaNO_3 + H_2O$$

131.5 g

$$= \quad \frac{131.5}{63} = 2.09 \text{ mol}$$

If 1 mol HNO_3 gives 1 mol $NaNO_3$,

then 2.09 mol HNO_3 gives 2.09 mol $NaNO_3$,

$$= \quad 2.09 \times 85 = 177.7 \text{ g } NaNO_3$$

The answer is that the yield of sodium nitrate is still 42.5 g. We cannot possibly get more than this, even though we have increased the amount of nitric acid, because there is no more sodium hydroxide with which it can react. We say that the nitric acid is *in excess*; the surplus merely remains in its original state and contaminates the sodium nitrate that is formed. The sodium hydroxide is said to be *deficient*, and therefore controls the yield. **Whenever two (or more) substances react together, and one of them is deficient, it is the deficient substance that controls the yield, and the calculation of yield must be based on this.**

If it is not immediately obvious which reactant is deficient, a simple calculation will decide.

Example

At room temperature, sulphuric acid and sodium chloride react together to give sodium hydrogensulphate and hydrogen chloride. Calculate the mass of gaseous hydrogen chloride produced when 73.0 g of pure sulphuric acid is allowed to react with 50.0 g of sodium chloride. (H = 1, O = 16, Na = 23, S = 32, Cl = 35.5)

Answer

$$H_2SO_4 + NaCl = NaHSO_4 + HCl$$

73.0 g	50.0 g
$= \dfrac{73.0}{98}$ mol	$\dfrac{50.0}{58.5}$ mol
$= 0.745$ mol	0.855 mol

The equation shows that equimolar quantities of sulphuric acid and sodium chloride are required; therefore 0.745 mol of sulphuric acid requires 0.745 mol of sodium chloride. This is referred to as the *theoretical requirement*. A comparison with the amount actually present, 0.855 mol, shows that sodium chloride is present in excess. There is a deficiency of sulphuric acid, and the calculation of yield must be based on the latter.

Since 1 mol H_2SO_4 gives 1 mol HCl,
0.745 mol H_2SO_4 gives 0.745 mol HCl,
$= 0.745 \times 36.5 = 27.2$ g HCl

Reactions involving gases

In the laboratory it is usually easier to measure the volume of a gas than its mass. Stoicheiometric calculations, however, involve moles. Fortunately, it is simple to convert the volume of a gas to an amount in moles (or vice versa) provided that we know the conditions under which the volume has been measured. The volumes of gases vary considerably with temperature and pressure, and must be compared under the same conditions. By convention, we adopt a standard temperature of 273 K and a standard pressure of 101.325 kPa (1 atm). These conditions are referred to as *standard temperature and pressure* (s.t.p.).

It is an experimental fact that **one mole of any gas at s.t.p. occupies approximately 22.4 dm³**. The foundation of this relationship lies in *Avogadro's law,* which states that equal volumes of all gases, under the same conditions of temperature and pressure, contain equal numbers of molecules. By inverting this statement we can argue that a given number of gas molecules (under fixed conditions) has a constant volume; therefore 6.02×10^{23} gas molecules, *i.e.* 1 mole (see p. 71), have a constant volume. By experiment, this is approximately 22.4 dm³ at s.t.p.

Example

On heating, mercury(II) oxide decomposes into mercury and oxygen. What volume of oxygen at s.t.p. can be obtained from 8.64 g of mercury(II) oxide? (O = 16, Hg = 200)

Answer

$2 \text{ HgO} = 2\text{Hg} + \text{O}_2$

8.64 g

$= \dfrac{8.64}{216} = 0.040 \text{ mol}$

From the equation,

2 mol HgO give 1 mol O_2,

therefore 0.040 mol HgO give 0.020 mol O_2

Now, 1 mol O_2 occupies 22.4 dm^3 at s.t.p.,

therefore 0.020 mol O_2 occupies $22.4 \times 0.020 = 0.0448$ dm^3 at s.t.p.

Often, gas volumes relate not to s.t.p. but to laboratory conditions. The conversion of a gas volume from laboratory conditions to s.t.p. or vice versa is carried out by means of the following relationship, which has been established by experimental work with many gases:

$$\frac{p_1 V_1}{T_1} = \frac{p_2 V_2}{T_2}$$

In this equation, which applies only to a fixed mass of gas, p stands for pressure, V for volume and T for temperature in kelvins, i.e. $^\circ$C + 273. The symbols on the left-hand side, each with a subscript 1, represent the initial condition of the gas, and those on the right, with subscript 2, the final condition.

Example

Calculate the volume of carbon dioxide at 290 K (17 $^\circ$C) and 98.4 kPa which is evolved when 50 g of calcium carbonate dissolves in excess hydrochloric acid. (H = 1, C = 12, O = 16, Cl = 35.5, Ca = 40)

Answer

The dissolving of calcium carbonate in hydrochloric acid is accompanied by a chemical reaction represented by the following equation:

$\text{CaCO}_3 + 2\text{HCl} = \text{CaCl}_2 + \text{CO}_2 + \text{H}_2\text{O}$

50 g

$= \dfrac{50}{100} = 0.5 \text{ mol}$

From the equation,

1 mol CaCO_3 gives 1 mol CO_2, i.e. 22.4 dm^3 CO_2 at s.t.p.,

therefore 0.5 mol CaCO_3 gives 11.2 dm^3 CO_2 at s.t.p.

To obtain the volume of carbon dioxide under the required conditions we use the equation quoted above. Initially, in our calculation, the gas is at s.t.p., so that $p_1 = 101$ kPa, $V_1 = 11.2$ dm^3 and $T_1 = 273$ K. In view of the final conditions we put $p_2 = 98.4$ kPa and $T_2 = 290$ K.

Therefore $\dfrac{101 \times 11.2}{273} = \dfrac{98.4 \times V_2}{290}$

therefore $V_2 = \dfrac{101 \times 11.2 \times 290}{273 \times 98.4} = 12.2 \text{ dm}^3 \text{ CO}_2$

Multi-stage reactions

When a calculation embraces more than one reaction, it is usually unnecessary to calculate the yields at intermediate stages.

Example

How much copper(II) sulphide can be obtained from 1.0 g of copper by the following series of reactions? (S = 32, Cu = 63.5)

i. Reacting the copper with dilute nitric acid:

$$3Cu + 8HNO_3 = 3Cu(NO_3)_2 + 2NO + 4H_2O$$

ii. Decomposing the copper(II) nitrate by heat:

$$2Cu(NO_3)_2 = 2CuO + 4NO_2 + O_2$$

iii. Reacting the copper(II) oxide with dilute sulphuric acid:

$$CuO + H_2SO_4 = CuSO_4 + H_2O$$

iv. Passing hydrogen sulphide gas through the solution:

$$CuSO_4 + H_2S = CuS + H_2SO_4$$

Answer

It is possible to calculate, first, the mass of copper(II) nitrate, and then in turn the masses of copper(II) oxide, copper(II) sulphate and copper(II) sulphide, but this is time-consuming and unnecessary. A study of the chemical equations shows that, provided there is an excess of reagents, one mole of metallic copper is eventually converted into one mole of copper(II) sulphide. No copper is lost in the form of other products.

We have 1.0 g of copper, *i.e.* $\dfrac{1.0}{63.5}$ mol of copper.

If 1 mol of Cu gives 1 mol of CuS,

then $\dfrac{1.0}{63.5}$ mol of Cu gives $\dfrac{1.0}{63.5}$ mol of CuS

$= \dfrac{1.0}{63.5} \times 95.5$ g of CuS $= 1.50$ g of CuS

Percentage yields

Whenever a chemical preparation is performed, the *actual yield, i.e.* the amount of product that is obtained, is always less than the theoretical yield (see p. 111). The main reasons for this are as follows.

The reaction may not proceed to completion

Many chemical reactions are said to be *irreversible* because the product (or products) is not easily reconverted into the reactants. Such reactions proceed to completion. An example is the burning of magnesium in oxygen to give magnesium oxide. Except at extremely high temperatures, magnesium oxide has no tendency to decompose into its elements. Consequently, all the magnesium is converted into its oxide.

However, there are some reactions which do not proceed to completion because they are *reversible*. In such cases the product (or products), under the conditions of the experiment, tends to revert to the original substances. An example is the combination of nitrogen and hydrogen to give ammonia. Ammonia is far less stable than the magnesium oxide in the previous example, and tends to decompose into nitrogen and hydrogen. For this reason, when nitrogen and hydrogen react together, a situation is eventually reached in which ammonia decomposes into its elements as fast as it is being formed. Thus, if we react one mole of nitrogen with three moles of hydrogen, according to the equation:

$$N_2 + 3H_2 = 2NH_3,$$

we cannot hope to obtain the theoretical yield of two moles of ammonia; only some amount less than two moles, depending on the experimental conditions, principally the temperature and the pressure.

Handling losses

Even if a reaction proceeds to completion, we can never achieve the theoretical yield in practice because some of the material always adheres to the apparatus in which it is made. A great deal can be done to minimise such losses, for example by scraping out solids with a spatula and by giving liquids time to drain completely. Much depends on the skill of the operator; clearly, someone who scatters material on the bench will not obtain a good yield!

One common operation which leads to handling losses is filtration (see p. 9). Both solids and liquids can be lost in this way, for the former tend to stick to the surface of the filter paper, while the latter are absorbed by the paper. The larger the paper the greater are the absorption losses, so it is important always to use a filter paper of the correct size.

With small scale preparations handling losses can be serious, for the waste of a few grams of material represents a high percentage loss.

Vaporisation losses

Volatile liquids and solids should always be kept in stoppered bottles or flasks. Stoppers not only minimise losses due to evaporation, but also reduce the fire risk in the case of flammable materials. In addition, they prevent contamination of the contents of the bottle by dust in the atmosphere, and prevent pollution of the atmosphere by the volatile liquid or solid.

Crystallisation losses

Dissolved solids are best recovered from solution by the technique of crystallisation (see p. 88) rather than by evaporation to dryness. In this way soluble impurities, provided they are present in relatively small proportions, stay in the mother liquor, *i.e.* the remaining solution. Unfortunately, some of the substance that is required also stays in solution and is lost.

Side reactions

When certain reactants are mixed together, more than one reaction can occur under a given set of conditions. The reactions are said to compete against one another. The reactions that occur to a small extent are known as *side reactions*, and because they consume some of the reactants they lead to a lowering of yield in the main reaction. There are many examples in organic chemistry, but relatively few in inorganic. One which is well known concerns the electrolysis of water (see p. 189). We may expect to obtain hydrogen at the cathode and oxygen at the anode in a volume ratio of exactly 2:1, because water decomposes according to the equation:

$$2H_2O = 2H_2 + O_2 .$$

In fact the yield of oxygen is slightly less than theoretical, because at the anode a side reaction occurs which leads to the formation of hydrogen peroxide, H_2O_2 :

$$2H_2O = H_2 + H_2O_2$$

An indication of the efficiency of a preparation is provided by the *percentage yield*, which is defined as follows:

$$\text{percentage yield} = \frac{\text{actual yield}}{\text{theoretical yield}} \times 100$$

The calculation should always relate to the purified product; never to crude material.

Determination of percentage yield in the preparation of calcium nitrate tetrahydrate

Calcium nitrate can be prepared by a chemical reaction between calcium hydroxide and nitric acid.

1. Weigh out on a preparative balance approximately 4 g of calcium hydroxide. Record the actual weight in your laboratory notebook.
2. Transfer the compound to an evaporating basin resting on a tripod and gauze, and add a few cubic centimetres of dilute nitric acid.
3. Warm the mixture, stir it with a glass rod, and then add further dilute nitric acid until a clear solution of calcium nitrate is obtained.
4. Boil the solution until the volume is reduced by about three-quarters; then allow it to cool down and crystallise. If it does not do so, continue the evaporation and again let the solution cool. Repeat the procedure if necessary until crystals are obtained. Some difficulty must be expected at this stage because calcium nitrate is highly soluble in water and does not crystallise well.
5. Filter off the crystals (see Chapter 2). Use a spatula to scrape as much material as possible from the evaporating basin into the filter paper.
6. Dry the crystals by pressing them between filter papers.
7. Weigh the product and record the actual yield.
8. Calculate the theoretical yield from the equation.

$$Ca(OH)_2 + 2HNO_3 + 2H_2O = Ca(NO_3)_2 .4H_2O$$

Relative molecular masses: 74   236

If x g of calcium hydroxide is weighed out, the amount of calcium hydroxide used is $\dfrac{x}{74}$ mol.

From the equation,

            1 mol $Ca(OH)_2$ gives 1   mol $Ca(NO_3)_2 .4H_2O$

therefore $\dfrac{x}{74}$ mol   ''   ''   '' $\dfrac{x}{74}$ mol   ''   ''   ''

            $= \dfrac{x}{74} \times 236$ g $Ca(NO_3)_2 .4H_2O$

9. Calculate the percentage yield, as defined above.

Factors affecting the rates of chemical reactions

The rate at which the reactants of a chemical reaction are converted into the products can vary enormously. At one extreme we have very rapid reactions between ions in solution, *e.g.*

$AgNO_3(aq) + NaCl(aq) = AgCl(s) + NaNO_3(aq)$
i.e. $Ag^+(aq) + Cl^-(aq) = AgCl(s)$

Immediately the solutions are mixed together, silver ions from the silver nitrate are attracted to chloride ions from the sodium chloride and combine to give a precipitate of silver chloride. (Although sodium ions are attracted to nitrate ions there is no precipitation of sodium nitrate because this compound is soluble in water.) Many other precipitation reactions are instantaneous for the same reason, namely that ions are present in solution at all times, and there is no delay while ions are being formed.

In contrast, certain geological reactions take place over hundreds or even thousands of years. An example is the weathering of granite, *i.e.* the slow chemical decomposition into china clay which occurs when granite is exposed to the atmosphere.

It is difficult to study the rates of very fast or very slow reactions in the laboratory. Between the extremes, however, there are many reactions that take place at a rate which is easily measured. Organic reactions commonly require a time of an hour or more, because covalent bonds have to be broken in order that the reactions can take place. Many of these covalent bonds, *e.g.* C–C, C–H and C–O, are strong and are not easily disrupted. The corrosion of a piece of iron may take several months or even several years, depending on the environment. Solids in general are relatively slow to react because chemical attack can occur only at the surface; the bulk of the solid is unable to take part.

The factors that can affect the rates of chemical reactions are as follows:
i.   temperature,
ii. concentration of reactants,
iii. catalysts,

iv.   pressure,

v.   surface area,

vi.   light.

The first three factors are particularly important because they have an effect on *all* reactions. We shall therefore discuss them first. Of the others, pressure affects only the rate of gas reactions, surface area is relevant only to solids, and light has an effect only on certain reactions involving covalent substances.

Temperature

If a certain reaction is slow at room temperature, our immediate instinct is to reach for a bunsen burner and heat up the reaction mixture. Every chemical reaction responds to a change of temperature. In general, the rate of a reaction is approximately doubled for a $10\ ^{\circ}C$ temperature rise, although in the case of very slow or very fast reactions the variation of reaction rate with temperature is scarcely noticeable.

To understand why an increase in temperature leads to an increase in the rate of reaction, we have to bear in mind that there are essentially two stages to any reaction, namely the breaking of bonds in the reactants and the formation of bonds in the products. We must concentrate on the first of these processes because it is the more difficult of the two and therefore controls the rate of reaction.

The breaking of bonds in the reactants is believed to occur when molecules collide with one another; but before discussing collisions between molecules it is useful to consider collisions between motor cars. A collision between two slow moving cars in a congested High Street is seldom serious, but a collision between two fast moving cars on a motorway may cause them to disintegrate. The cars will not *necessarily* break up; a great deal depends on the manner in which they approach each other, for a glancing blow will not cause much damage. Now the *kinetic theory* tells us that molecules of gases and liquids move about continuously, with an average velocity that depends on the temperature. They travel in a much more random fashion than motor cars, but like cars they collide with one another and these collisions *may* lead to disintegration. At low temperatures, where the average molecular velocity is low, relatively few collisions lead to the breaking of covalent bonds and the rate of reaction is low. At higher temperatures, however, molecular velocities are greater and more of the collisions occur with the necessary impact to break covalent bonds. The rate of reaction is therefore increased.

Concentration of reactants

Metals, such as magnesium, are often slow to dissolve in very dilute hydrochloric acid. However, if more concentrated acid is used the same reaction still occurs, namely:

$$Mg + 2HCl = MgCl_2 + H_2,$$

but at an increased rate.

We can easily investigate the relationship between the rate of this reaction and the concentration of hydrochloric acid. Into a number of beakers we pour hydrochloric acid of various concentrations, *e.g.* 4 mol dm^{-3} (4 M) in the first, 2 mol dm^{-3} (2 M) in the second, then, perhaps, 1 mol dm^{-3}, 0.5 mol dm^{-3} and 0.25 mol dm^{-3} (1 M, 0.5 M and 0.25 M respectively) in the others. To each beaker we add the same length (and hence the same mass) of magnesium ribbon, and note the time (t) for the magnesium to dissolve in each case. It is important, particularly with very dilute acid, to ensure that the acid is present in excess, and to stir the solution to avoid the acid being completely used up in the vicinity of the magnesium.

The rate of this reaction can be represented by $1/t$, and if we plot a graph of $1/t$ against concentration of the acid we obtain a straight line through the origin. From this we conclude that the rate of the reaction is directly related to the concentration of the hydrochloric acid.

Experiments have shown that the rate of any reaction which occurs in solution depends on the concentration of each of the reacting substances. This is a simplified statement of the *kinetic law of mass action,* which is one of the most important laws of modern chemistry. For a simple reaction represented by the equation:

A = B + C,
rate $\propto$ [A],

where [A] represents the concentration of A. Square brackets are often used in chemistry to denote concentrations in mol dm^{-3} (molarity). Thus, if we double the concentration of A, we double the rate of reaction; if we treble [A], we treble the rate, and so on.

For a reaction between one molecule of A and one of B,

A + B = C + D,

the rate is proportional to the concentrations of both A and B,

i.e. rate $\propto$ [A] [B]

Thus, if we double the concentration of A we double the rate of reaction, and if we double the concentration of B we also double the rate of reaction. If we double the concentration of A *and* double the concentration of B we quadruple the original rate of reaction.

If a reaction is represented by the equation:

2A = B + C,
then rate $\propto$ [A]^2

This is because the equation can be written as

A + A = B + C,
in which case rate $\propto$ [A] [A]

For a general case,

aA + bB = products,
rate $\propto$ [A]^a[B]^b

It will be noticed that the nature of the products never has any influence on the rates of reactions in solution.

Great care must be taken to apply the kinetic law of mass action only to single reactions. Many chemical changes take place not by one reaction, but by two or more reactions which occur one after the other, often in rapid succession. In such cases the kinetic law of mass action applies to any reaction in the series, but not to the series as a whole. For example, suppose a reaction:

$$A + B = C \tag{1}$$

is immediately followed by:

$$A + C = D + E \tag{2}$$

so that the complete change is represented by an equation obtained by adding together [1] and [2]:

$$2A + B = D + E \tag{3}$$

Then we can apply the kinetic law of mass action to [1] and [2], but not to [3]. In other words,

rate of reaction [1] $\propto$ [A] [B],
and rate of reaction [2] $\propto$ [A] [C],

but the rate of the complete change [3] is *not* $\propto$ $[A]^2$ [B].

The dependence of reaction rate upon concentration is readily explained by the kinetic theory. Whenever the molar concentration of a reactant is low, the *molecular* concentration is low, *i.e.* molecules are relatively far apart from one another, and the likelihood of collisions between them is relatively small. At higher concentrations, however, the molecules are closer together; the chances of collision are greater, and the rate of reaction is greater.

Catalysts

The rates of chemical reactions can be affected by the presence, often in small amounts, of substances called *catalysts*. A catalyst can be recovered, unchanged chemically and unchanged in mass, at the end of a reaction, and appears to take no part in the reaction. Appearances, however, are deceptive, and we shall see that a catalyst plays a very real part in the way that a reaction takes place. If a catalyst did not interact in some way with at least one of the reactants, there would be no point in including it.

Catalysts which accelerate reactions are called *positive catalysts* or *accelerators*, while those which retard reactions are called *negative catalysts* or *inhibitors*. Consider, for example, the decomposition of aqueous hydrogen peroxide into water and oxygen:

$$2H_2O_2 = 2H_2O + O_2$$

The decomposition, normally slow at room temperature, can be greatly assisted by the addition of a small amount of manganese(IV) oxide, which acts as a positive catalyst. In contrast, the decomposition can be slowed down by the inclusion of a negative catalyst such as ethanol or glycerol.

As a general rule, catalysts are not interchangeable. Consider the following reactions:

$$N_2 + 3H_2 = 2NH_3$$
$$2KClO_3 = 2KCl + 3O_2$$

The former, which is the basis of the *Haber process* for the manufacture of ammonia, can be catalysed by iron. The latter reaction, which is often used for preparing limited amounts of oxygen in school laboratories, is catalysed by manganese(IV) oxide. These catalysts are useless if they are interchanged. Iron is specific for the first reaction and certain other gas reactions. Manganese (IV) oxide will catalyse only the thermal decomposition of potassium chlorate and a few other reactions, such as the decomposition of hydrogen peroxide quoted above.

The most specific catalysts of all are *enzymes*. Enzymes are organic substances called proteins, and they occur in the cells of plants and animals. The function of an enzyme is to catalyse a particular reaction in the cell where it occurs, and in general it will do nothing but that. Well known examples of enzyme catalysis are as follows:

$$2H_2O_2 \xrightarrow[\text{from blood or milk}]{\text{enzyme 'catalase'}} 2H_2O + O_2$$

$$C_6H_{12}O_6 \xrightarrow[\text{from yeast}]{\text{enzyme 'zymase'}} 2C_2H_5OH + 2CO_2$$
glucose $\qquad\qquad\qquad\qquad$ ethanol

Enzymes can be used to catalyse reactions in the laboratory and in industry. The brewing industry, for example, relies on the fermentation of glucose according to the second equation shown above.

Two general types of catalysis are recognised. In one type the catalyst is in solution with the reactants, while in the other the catalyst is present as a solid in the reacting liquids or gases.

An example of the first type is the conversion of iodide ions, I^-, (*e.g.* from potassium iodide) to iodine, I_2, by means of peroxodisulphate ions, $S_2O_8^{2-}$, (*e.g.* from ammonium peroxodisulphate):

$$2I^- + S_2O_8^{2-} = I_2 + 2SO_4^{2-}$$

The reaction is accelerated by iron(II) sulphate, $FeSO_4$. The catalyst functions by providing an easy route for the reaction. Iron(II) ions, Fe^{2+}, from the $FeSO_4$ react with the peroxodisulphate ions to form sulphate ions and an *intermediate*, namely iron(III) ions, Fe^{3+}:

$$2Fe^{2+} + S_2O_8^{2-} = 2Fe^{3+} + 2SO_4^{2-}$$
catalyst $\quad$ reactant $\quad$ intermediate $\quad$ product

The intermediate then reacts with iodide ions to produce iodine. At the same time, iron(II) ions are reformed and are able to bring about further change:

$$\underset{\text{intermediate}}{Fe^{3+}} \quad + \quad \underset{\text{reactant}}{2I^-} \quad = \quad \underset{\text{product}}{I_2} \quad + \quad \underset{\text{catalyst}}{Fe^{2+}}$$

A catalyst is always regenerated at the end of a catalysed reaction.

An example of the second type of catalysis is the formation of ammonia by the combination of (gaseous) nitrogen and hydrogen in the presence of (solid) platinum or iron. The catalyst again functions by providing an easy route for the reaction, but in this case it does so by *adsorbing* the reactants, nitrogen and hydrogen, on to its surface. 'A*d*sorption' means the retention of gas on the surface of a solid. The gas does not penetrate the bulk of the solid. In contrast, a sponge is said to 'a*b*sorb' water, because the whole of the sponge is involved.

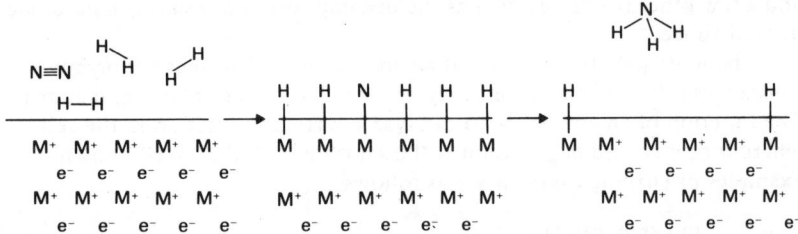

(a) Metal + molecules of nitrogen and hydrogen   (b) Metal + adsorbed gas   (c) Release of a molecule of ammonia

Fig. 9.2 The combination, on the surface of a metal catalyst, of nitrogen and hydrogen to give ammonia (M = metal)

In the adsorbed state (see Fig. 9.2(b)), atoms of nitrogen and hydrogen are much closer together than are molecules in the gaseous state, and the elements are much more likely to combine to produce ammonia. Ammonia molecules, once formed, escape from the surface of the metal, leaving empty spaces where further adsorption and reaction can occur (see Fig. 9.2(c)).

Industrial applications

Economic considerations dictate the wide use of catalysts in the chemical industry. Well chosen catalysts enable reactions to be carried out at relatively low temperatures, and also reduce the time required for reactions to occur. Some well known industrial examples of catalysed reactions are shown in Table 9.1.

Pressure

An increase of pressure increases the rate of any gas reaction, by forcing the reactant molecules closer together. An increase of pressure, for gas reactions, is thus analogous to an increase of concentration for reactions which occur in solution. In both cases there is an increase in the number of molecules per unit volume, which leads to an increase in the number of collisions and an increase in reaction rate.

Table 9.1 Catalysts of industrial importance

Process	Equation	Catalyst
Contact process	$2SO_2 + O_2 = 2SO_3$	V_2O_5
Haber process	$N_2 + 3H_2 = 2NH_3$	Fe
Methanol manufacture	$CO + 2H_2 = CH_3OH$	$Cr_2O_3 + ZnO$
Polythene manufacture	$nCH_2{=}CH_2 = (-CH_2-CH_2-)_n$	Titanium compounds
Margarine manufacture	$\underset{\substack{\\ \text{in soya}\\ \text{bean oil}}}{\overset{\diagdown}{\diagup}C{=}C\overset{\diagup}{\diagdown}} + H_2 = -\overset{\overset{\displaystyle H}{\mid}}{C}-\overset{\overset{\displaystyle H}{\mid}}{C}-$	Ni

An increase of pressure has no influence on the rate at which solids and liquids react, because their molecules cannot be forced closer together.

Surface area

Solids can react only at their surfaces. It follows that the rate at which a solid reacts with a liquid or a gas depends on its surface area. This can readily be demonstrated by comparing the times taken for an excess of hydrochloric acid to dissolve a given mass of aluminium in the form of (i) a lump, (ii) foil, and (iii) powder. The lump, which presents the smallest surface area to the acid, dissolves slowly; the foil, which has a greater surface area, rather more rapidly; and the powder, which exposes the largest surface area to attack, is the fastest of all to dissolve.

The finer the powder, the faster is the reaction. This is because subdivision greatly increases the surface area, as we can see by considering a sphere of radius (r) 1 mm.

$$\text{Surface area of the sphere} = 4\pi r^2 = 4 \times \frac{22}{7} \times 1^2 = 12.6 \text{ mm}^2$$

If we cut the sphere in two, each hemisphere has a surface area of ½ × 12.6 mm^2, plus the area of the circle (πr^2) formed on bisection, *i.e.* area of hemisphere $= 6.3 + (\frac{22}{7} \times 1^2) = 9.4 \text{ mm}^2$

therefore total surface area after cutting $= 2 \times 9.4 = 18.8 \text{ mm}^2$

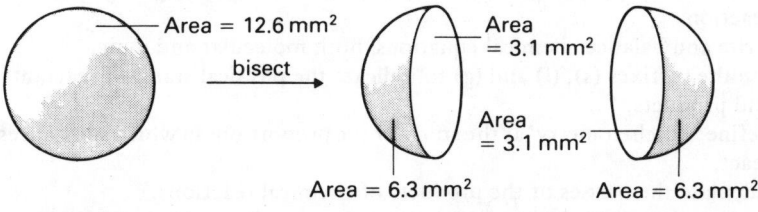

Sphere   Two hemispheres

Fig. 9.3 The increase of surface area on bisecting a sphere

Light

Ultraviolet light, and also visible light at the violet end of the spectrum, has the ability to break certain types of covalent bond and hence bring about some chemical changes. For example, hydrogen and chlorine do not combine in the dark at room temperature, but will do so explosively, to give hydrogen chloride, on exposure to ultraviolet light:

$$H_2 + Cl_2 = 2HCl$$

The reason is that ultraviolet light is able to break the covalent bond in some of the chlorine molecules to give chlorine atoms:

$$Cl_2 = 2Cl$$

Chlorine atoms then combine with molecules of hydrogen to give molecules of hydrogen chloride and atoms of hydrogen:

$$Cl + H_2 = HCl + H$$

Hydrogen atoms attack molecules of chlorine in a similar fashion:

$$H + Cl_2 = HCl + Cl$$

These two reactions alternate with each other until the concentration of either hydrogen or chlorine falls to a low level.

The reason why plants cannot thrive in the dark is that they require sugars for their growth, and these compounds are formed in the leaves only in the presence of light. The reaction, known as *photosynthesis*, involves the combination of carbon dioxide and water:

$$6CO_2 + 6H_2O = C_6H_{12}O_6 + 6O_2$$

The green colouring material in the leaves, called chlorophyll, acts as a catalyst for this reaction.

Summary

At the conclusion of this chapter, you should be able to:
1. recognise that chemical change consists of one or more chemical reactions,
2. state the Law of Conservation of Mass,
3. recognise that chemical reactions may be exothermic or endothermic,
4. define 'enthalpy change' as the heat change accompanying a chemical reaction,
5. write and balance chemical equations, both molecular and ionic,
6. use the suffixes (s), (l) and (g) to indicate the physical states of reactants and products,
7. define 'stoicheiometry' as the study of the proportions in which substances react,
8. calculate the masses of the products of chemical reactions,
9. state the conditions of standard temperature and pressure (s.t.p.),
10 state that one mole of any gas at s.t.p. occupies approximately 22.4 dm^3,

11. state the gas equation, $\dfrac{p_1 V_1}{T_1} = \dfrac{p_2 V_2}{T_2}$,

12. calculate the volumes of gaseous products of chemical reactions,
13. calculate the percentage yields of chemical preparations,
14. perform experiments to investigate the stoicheiometry of reactions,
15. state that the rate of a reaction generally increases considerably with an increase of temperature,
16. state that reaction rate in solution increases with the concentration of the reactants,
17. define a 'catalyst' as a substance which influences the rate of a reaction, while itself remaining chemically unchanged at the end of the reaction,
18. state examples of catalysts of industrial importance,
19. state that the rate of a gas reaction increases with an increase of pressure,
20. state that the rate at which a solid reacts depends on its degree of sub-division.

Questions

1. Write balanced stoicheiometric (*i.e.* molecular) equations for the following reactions:
   (*a*) the thermal decomposition of copper(II) carbonate into copper(II) oxide and carbon dioxide,
   (*b*) the thermal decomposition of sodium hydrogencarbonate into sodium carbonate, carbon dioxide and water,
   (*c*) the combination of iron and chlorine to give iron(III) chloride,
   (*d*) the combination of aluminium and oxygen to give aluminium oxide,
   (*e*) the dissolving of calcium in water to give calcium hydroxide and hydrogen,
   (*f*) the dissolving of aluminium in dilute sulphuric acid to give aluminium sulphate and hydrogen.

   In questions 2–8 the following relative atomic masses are required:
   H = 1, N = 14, O = 16, Na = 23, Al = 27, S = 32, Cl = 35.5, Fe = 56, Ni = 59, Zn = 65.

2. Calculate the mass of sodium nitrite, $NaNO_2$, that can be obtained by heating 1.70 g of sodium nitrate.

   $$2NaNO_3 = 2NaNO_2 + O_2$$

3. Calculate the mass of aluminium hydroxide that must be heated to obtain 10.00 g of aluminium oxide.

   $$2Al(OH)_3 = Al_2O_3 + 3H_2O$$

4. Calculate the mass of anhydrous zinc sulphate that can be obtained by dissolving 65.00 g of zinc in:
   (*a*) excess dilute sulphuric acid,
   (*b*) dilute sulphuric acid made by dissolving 90.00 g of the pure acid in water.

   $$Zn + H_2SO_4 = ZnSO_4 + H_2$$

5. If 8.00 g of sodium hydroxide is heated with 10.70 g of ammonium chloride, calculate:
   (a) the maximum mass of ammonia that can be obtained,
   (b) the volume of gaseous ammonia at s.t.p.,
   (c) the volume of gaseous ammonia at 20 °C (293 K) and 95 kPa (0.95 atm).

   $$NaOH + NH_4 Cl = NaCl + NH_3 + H_2 O$$

6. Calculate the mass of steam produced when 10 dm^3 of hydrogen is exploded with 4 dm^3 of oxygen. (Both volumes relate to s.t.p.)

   $$2H_2 + O_2 = 2H_2 O$$

7. Calculate the maximum mass of nickel(II) sulphide that can be precipitated by passing 0.5 dm^3 of gaseous hydrogen sulphide, measured at 15 °C (288 K) and 98.7 kPa (0.987 atm), through a solution of excess nickel(II) sulphate.

   $$NiSO_4 + H_2 S = NiS + H_2 SO_4$$

8. Copper(II) sulphate and potassium iodide react together to give copper(I) iodide and iodine. The liberated iodine can be reacted with sodium thiosulphate, $Na_2 S_2 O_3$, to give sodium tetrathionate, $Na_2 S_4 O_6$, and sodium iodide, NaI.

   $$2CuSO_4 + 4KI = 2CuI + I_2 + 2K_2 SO_4$$

   $$I_2 + 2Na_2 S_2 O_3 = Na_2 S_4 O_6 + 2NaI$$

   Assuming that the potassium iodide is present in excess, calculate the number of moles of sodium thiosulphate required to react with the iodine liberated by one mole of copper(II) sulphate.

   In questions 9–13 select the most appropriate answer, labelled A, B, C or D.

9. Which one of the following factors does *not* influence the rate at which zinc dissolves in dilute hydrochloric acid?
   A   Temperature.
   B   Pressure.
   C   Concentration of hydrochloric acid.
   D   State of subdivision of zinc.

10. The rate of the reaction between gaseous nitrogen and gaseous hydrogen can be reduced by:
    A   introducing an iron catalyst,
    B   raising the temperature,
    C   reducing the pressure,
    D   exposing the mixture to ultraviolet light.

11. The most suitable catalyst for the contact process for the manufacture of

sulphuric acid is:

A   iron,

B   nickel,

C   vanadium(V) oxide,

D   titanium compounds.

12. Which of the following statements about catalysts is *untrue*?

A   A catalyst can be recovered, chemically unchanged, at the end of a reaction.

B   A catalyst plays no part in a reaction.

C   At the end of a reaction, the mass of a catalyst is equal to its original mass.

D   A positive catalyst provides a relatively easy route for a reaction to occur.

13. The main reason why an increase in temperature leads to an increase in rate of reaction is that:

A   internal strains develop inside molecules,

B   intermolecular attraction increases,

C   the number of intermolecular collisions increases,

D   intermolecular collisions occur with more force.

Chapter 10

Chemical properties of the elements

In Chapter 4 we studied the physical characteristics of the elements, both metals and non-metals. Here we shall consider some of their chemical properties, namely combination with oxygen, combination with hydrogen and reaction with water.

The reactions between elements and acids will be discussed in Chapter 12.

Combination with oxygen

Some elements, notably sodium and calcium, combine rapidly with oxygen to form compounds called *oxides*. Other elements, such as magnesium, sulphur or carbon, form oxides only when heated. Such reactions are conveniently carried out by heating the element concerned in air, which contains about 20 per cent of oxygen. Atmospheric nitrogen is generally not involved in these reactions.

Often, when an element is heated in air, heat and light are given out as the reaction with oxygen takes place. In such cases we say that the element is 'burning' or that *combustion* is occurring. Thus, the combustion of many elements in the atmosphere involves combination with oxygen to form oxides.

The nature of oxides

A distinctive feature of all metals is that they each form at least one oxide which has basic properties, while non-metal oxides are acidic.

A *basic oxide* will react with and neutralise an acid to form a salt and water, *e.g.*

$$MgO \quad + \quad 2HCl \quad = \quad MgCl_2 \quad + \quad H_2O$$
magnesium oxide $\qquad\qquad\qquad$ magnesium chloride

$$Fe_2O_3 \quad + \quad 3H_2SO_4 \quad = \quad Fe_2(SO_4)_3 \quad + \quad 3H_2O$$
iron(III) oxide $\qquad\qquad\qquad$ iron(III) sulphate

Most *acidic oxides* react with water to produce acids, *e.g.*

$$H_2O + SO_3 \quad = H_2SO_4 \quad \text{sulphuric acid}$$
$$6H_2O + P_4O_{10} = 4H_3PO_4 \quad \text{phosphoric acid}$$

In addition, an acidic oxide will combine with a basic oxide to form a salt *only*, *e.g.*

$$CaO + SO_3 \quad = CaSO_4 \quad \text{calcium sulphate}$$
$$Na_2O + CO_2 = Na_2CO_3 \quad \text{sodium carbonate}$$

Some non-metal oxides, notably carbon monoxide, give neutral solutions in water and react with bases only under drastic conditions. They are therefore said to be 'weakly acidic'.

The reactions of some elements with oxygen

Magnesium

Magnesium reacts very slowly with oxygen at room temperature to form white magnesium oxide, MgO. At higher temperatures the reaction is extremely rapid and is accompanied by the evolution of much heat and light, *i.e.* heated magnesium burns in oxygen. It is important not to look directly at burning magnesium because the intense light can damage your eyesight. The product of combustion is again magnesium oxide, but in air a small amount of magnesium nitride, Mg_3N_2, is also formed. The fact that magnesium oxide is basic (see above) supports the classification of magnesium as a metallic element.

Iron

Dry air or oxygen does not react appreciably with iron at room temperature. However, when a piece of iron is heated in air or oxygen the surface of the iron loses its bright appearance and becomes covered with a layer of oxides of iron, which varies in colour from brown to black. If a piece of thin iron wire (or iron filings) is heated to red heat (approximately 600 °C (873 K)) in oxygen a spectacular reaction occurs which looks very much like a 'sparkler' burning on Guy Fawkes night. This is another example of combustion. The product is a black oxide of iron, Fe_3O_4, called triiron tetraoxide:

$$3Fe + 2O_2 = Fe_3O_4$$

Triiron tetraoxide is a basic oxide, which agrees with the classification of iron as a metal.

In the presence of water, iron is attacked fairly rapidly by oxygen at room temperature to give that familiar material, 'rust'. A number of reactions take place in the rusting process, and the final product is $Fe_2O_3(aq)$, *i.e.* iron (III) oxide with some water molecules bound to it.

Copper

Like iron, copper is not attacked by dry air or oxygen at room temperature. Heated copper combines fairly readily with oxygen to form black copper(II) oxide, CuO, which is again a basic oxide:

$$2Cu + O_2 = 2CuO$$

In the presence of water the reaction between copper and oxygen is more complicated.

Hydrogen

Hydrogen burns in oxygen to give water:

$$2H_2 + O_2 = 2H_2O$$

Because of this, water can be regarded as the oxide of hydrogen.

Water is not a typical non-metallic oxide because it does not react with bases to give salts. It is unique in that it is a 'neutral oxide', with no obvious acidic or basic character.

Sulphur

Sulphur does not react with oxygen at room temperature. If sulphur is heated in air, or if hot sulphur is placed in oxygen, the sulphur burns with a small blue flame and white smoke is produced. This is another example of combustion. The main product of the reaction is sulphur dioxide, a strongly smelling, poisonous gas:

$$S + O_2 = SO_2$$

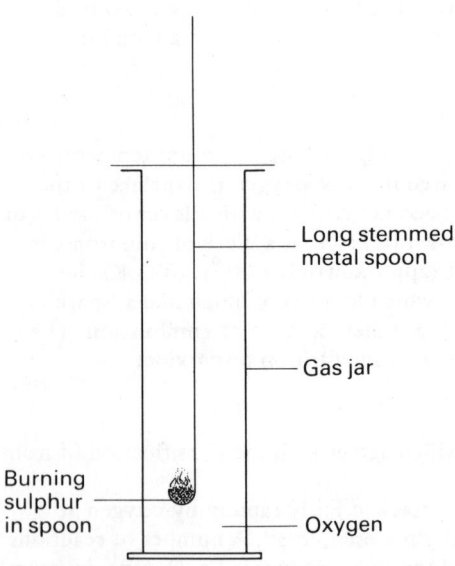

Long stemmed metal spoon

Gas jar

Burning sulphur in spoon

Oxygen

Fig. 10.1 Sulphur burning in a gas jar of oxygen

A little solid sulphur trioxide is also formed:

$$2S + 3O_2 = 2SO_3$$

The presence of small solid particles of sulphur trioxide in the gaseous sulphur dioxide are responsible for the smoky appearance.

Because sulphur is a non-metal, both these oxides are acidic.

Carbon

Graphite and diamond react with oxygen only when heated to temperatures greater than 700 $^\circ$C (973 K). In the presence of a large amount of oxygen combustion occurs and carbon dioxide, CO_2, is produced:

$$C + O_2 = CO_2$$

With a smaller amount of oxygen carbon monoxide, CO, is also formed:

$$2C + O_2 = 2CO$$

Both of these oxides are acidic, in accordance with the classification of carbon as a non-metal.

Oxidation and reduction

A reaction in which a substance combines with oxygen is referred to as *oxidation*. All the reactions described above are examples of oxidation, because in each case an element combines with oxygen to form an oxide.

The removal of oxygen from the oxide of a metal results in the metal being reformed. This process, which is the opposite of oxidation, is known as *reduction*. The term 'reduction' originally meant 'reduction in oxygen content', and is used to describe any reaction in which oxygen is removed from a substance.

Thus, we can say that copper is *oxidised* to form copper(II) oxide in an oxidation reaction. The reverse change is a reduction, in which copper(II) oxide is said to be *reduced* to copper.

The safe use of oxygen cylinders

The oxygen required for combustion and other oxidation reactions is normally provided by the atmosphere. If, however, pure oxygen is required, it can conveniently be obtained from a gas cylinder. All gas cylinders are a distinctive colour. Oxygen cylinders can be identified because they are painted black, while those containing compressed air are grey.

A potentially hazardous situation can arise when oxygen cylinders are in use. This is because the surrounding air becomes enriched in oxygen, *i.e.* the concentration of oxygen rises above its normal value of about 21 per cent. Under these conditions flammable substances may ignite more easily than usual. It is also possible for *spontaneous combustion* to occur, *i.e.* substances can catch fire of their own accord, without the application of a flame.

Such a situation is most likely to arise if oxygen is used in a confined space, such as a fume cupboard. The danger can be minimised by a good fume extraction system. **Another safety measure is the avoidance of grease on the**

thread of the pressure regulator valve on top of the cylinder. This precaution applies to gas cylinders in general.

Combination with hydrogen

Most elements, both metals and non-metals, combine with hydrogen to form compounds called *hydrides*. In general, combination with hydrogen occurs much less rapidly than with oxygen, and most hydrides are formed only on heating. For example, hydrogen and nitrogen combine to give ammonia, NH_3, at a temperature of 500 °C (773 K) and a pressure of 200 atm (20 000 kPa) in the presence of iron as a catalyst:

$$N_2 + 3H_2 = 2NH_3$$

However, a mixture of hydrogen and chlorine will combine explosively to give hydrogen chloride if the mixture of gases is exposed to sunlight or if an electric spark is passed through it:

$$H_2 + Cl_2 = 2HCl$$

Sodium combines with hydrogen at a temperature of 380 °C (653 K) to give a white solid called sodium hydride:

$$2Na + H_2 = 2NaH$$

The reaction is accompanied by the evolution of light and heat. We can say, therefore, that sodium undergoes combustion in hydrogen as well as in oxygen. It is one of the few elements to do so.

Some elements, such as carbon, are reluctant to form hydrides by direct combination of the elements.

The nature of hydrides

The alkali metals (*e.g.* sodium) and the alkaline earth metals (*e.g.* calcium) form hydrides which, like the corresponding oxides, are basic in character. Most other metals form hydrides which are highly unreactive and in which this basic character is totally absent.

The nature of non-metallic hydrides varies greatly. Ammonia, for example, is basic, but hydrogen chloride is acidic. Methane and other hydrocarbons, *i.e.* the hydrides of carbon, have neither basic nor acidic properties and can be described as 'neutral'.

Oxidation and reduction

The addition or removal of oxygen is not the only way in which oxidation or reduction can occur. The addition of hydrogen to a substance is also termed 'reduction', while the removal of hydrogen is 'oxidation'.

For example, in the reaction between nitrogen and hydrogen to give ammonia, the nitrogen is said to be reduced by the addition of hydrogen. The reconversion of ammonia to nitrogen involves the loss of hydrogen and can be described as an oxidation reaction.

The formation of metals from their oxides by reduction with hydrogen

The oxides of reactive metals, such as sodium, potassium, calcium and magnesium, are extremely difficult to reduce. However, those of many other metals, notably iron, copper, lead and silver, are reduced much more easily. To bring about these reactions the presence of another substance is necessary in order to combine with, and hence remove, the oxygen of the metal oxide. Such a substance is called a *reducing agent*. Common reducing agents for metal oxides are hydrogen, carbon or carbon monoxide. Here we shall consider the use of hydrogen only. When the oxides of iron, copper, lead or silver are heated in a stream of hydrogen in a combustion tube (see Fig. 10.2) they are reduced to the metal, *e.g.*

$CuO + H_2 = Cu + H_2O$
$PbO + H_2 = Pb + H_2O$

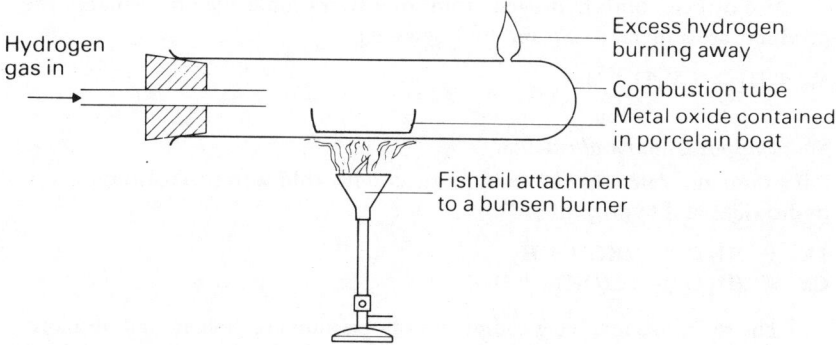

Fig. 10.2 Apparatus for the reduction of metal oxides

The safe use of hydrogen cylinders

The hydrogen for reduction experiments is conveniently obtained from a gas cylinder. A hydrogen cylinder is painted red for identification, and is equipped with a pressure regulator valve which is screwed into the top of the cylinder by a left-handed thread. This is in contrast to cylinders containing oxygen, nitrogen or compressed air, all of which have a right-handed thread.

Great care must always be exercised when using this gas, for mixtures of hydrogen and air explode violently when heated or sparked. Thus, when reducing a metal oxide with hydrogen, **it is very important, before heating is commenced, to pass hydrogen through the apparatus for several minutes to ensure that all the air is removed.**

Excess hydrogen which passes out of the apparatus should be burnt **by the lecturer in charge** or vented into a fume cupboard and not allowed to escape into the atmosphere. Hydrogen is less dense than air, and rapidly rises to the ceiling of a laboratory. If the gas is not swept away by draughts it collects and forms a flammable mixture with air. Any sparking that may occur in ceiling lights, particularly the fluorescent variety when they are switched on, can cause the mixture to explode.

All connections to a hydrogen cylinder should periodically be checked by putting a few drops of a detergent solution on them. Any leaks show up as a fine stream of bubbles.

Reaction with water

Metals

Relatively few of the common metals react with water in the absence of oxygen or other reagents. When a metal does react with water, one or both of the hydrogen atoms from a water molecule are displaced by the metal. If only one hydrogen atom is displaced then the metal hydroxide and hydrogen are produced, *e.g.*

$$2Na + 2H_2O = 2NaOH + H_2$$

If, however, both hydrogen atoms of a water molecule are displaced the products are a metal *oxide* and hydrogen, *e.g.*

$$Mg + H_2O = MgO + H_2$$

Sodium, potassium and calcium

Potassium and calcium, like sodium, react with cold water to form their hydroxides and hydrogen:

$$2K + 2H_2O = 2KOH + H_2$$
$$Ca + 2H_2O = Ca(OH)_2 + H_2$$

The reactions involving sodium and potassium are violent, and no more than small pellets of either metal should be used. Sufficient heat is given out in the reactions to melt the metals so that they form tiny molten globules which skate around the surface of the water. In the case of potassium so much heat is evolved that the hydrogen catches fire. After the reaction, a clear solution of sodium hydroxide or potassium hydroxide remains.

The reaction of calcium with water is much less vigorous than that with sodium or potassium. The mixture goes cloudy as the reaction proceeds because sparingly soluble calcium hydroxide is formed.

Magnesium and zinc

Cold water does not appreciably attack these metals. When heated in steam, however, the metals burn to produce their oxides. Two atoms of hydrogen per water molecule are therefore displaced by the metal:

$$Zn + H_2O = ZnO + H_2$$

At the end of the reaction a white oxide, MgO or ZnO, remains.

Aluminium

The surface of a piece of aluminium is always covered by a thin, almost invisible film of aluminium oxide, Al_2O_3, which prevents water from coming into contact with the metal itself. If the film is removed or loosened then reaction occurs slowly to give aluminium hydroxide and hydrogen:

$$2Al + 6H_2O = 2Al(OH)_3 + 3H_2$$

The oxide film may be removed or loosened by scrubbing the aluminium with wire wool, or by rubbing it with mercury.

Iron

Cold water, in the absence of air, does not react with iron. At approximately 500 °C (773 K), iron reacts with steam in a reversible reaction (see p. 117) to form triiron tetraoxide and hydrogen:

$$3Fe + 4H_2O = Fe_3O_4 + 4H_2$$

Other metals

The remaining common metals, *e.g.* tin, lead, copper, mercury, silver and gold, do not react with water even when heated in steam.

Non-metals

The only non-metals which react with water are the halogens. Chlorine, for example, dissolves in water to give a solution, called 'chlorine water', which contains hydrochloric acid, HCl, and hypochlorous acid, HClO. Both acids arise through a chemical reaction between the chlorine and the water:

$$Cl_2 + H_2O = HCl + HClO$$

On standing, chlorine water gives off oxygen as the hypochlorous acid decomposes:

$$2HClO = 2HCl + O_2$$

The complete chemical change can be represented by a single equation obtained by combining these two:

$$2Cl_2 + 2H_2O = 4HCl + O_2$$

Summary

At the conclusion of this chapter, you should be able to:
1. state that combustion in air involves combination with oxygen,
2. recognise that, in general, metal oxides are basic and non-metal oxides are acidic,
3. describe the reactions of some common elements with oxygen and the nature of the resulting oxides,
4. describe water as an oxide of hydrogen,
5. describe combination with oxygen as 'oxidation' and removal of oxygen as 'reduction',
6. state the hazards associated with the use of oxygen cylinders,
7. describe the reactions of some common elements with hydrogen and the nature of the resulting hydrides,

8. describe addition of hydrogen as 'reduction' and removal of hydrogen as 'oxidation',
9. describe the reactions of metal oxides with hydrogen,
10. state the hazards associated with hydrogen and hydrogen cylinders,
11. state the colour codes for cylinders of oxygen, compressed air and hydrogen,
12. describe the reactions of some common elements with water.

Questions

1. State which of the following oxides would probably be basic and which would probably be acidic.

   (a) SnO   (e) Cl_2O
   (b) P_4O_6 (f) Ag_2O
   (c) Na_2O (g) CaO
   (d) HgO   (h) NO_2

2. Give the formula of the oxide formed by heating each of the following elements in oxygen, and state whether it is basic, acidic or neutral.

   (a) Iron   (e) Phosphorus
   (b) Calcium   (f) Copper
   (c) Hydrogen   (g) Carbon
   (d) Sulphur   (h) Sodium

   In questions 3–8, select the most appropriate answer, labelled A, B, C or D.

3. In which one of the following reactions is the oxide being oxidised?
   A $CuO + C = Cu + CO$
   B $2BaO_2 = 2BaO + O_2$
   C $2CO + O_2 = 2CO_2$
   D $2SO_3 = 2SO_2 + O_2$

4. Which one of the following gas cylinders is painted red?
   A   Hydrogen.
   B   Oxygen.
   C   Compressed air.
   D   Nitrogen.

5. In which one of the following equations is the first named species not acting as a reducing agent?
   A $H_2 + CuO = Cu + H_2O$
   B $C + PbO = Pb + CO$
   C $3CO + Fe_2O_3 = 2Fe + 3CO_2$
   D $NO_2 + SO_2 = NO + SO_3$

6. Which one of the following metals reacts with water or steam to produce an oxide?
   A   Magnesium.
   B   Potassium.
   C   Calcium.
   D   Sodium.

7. Which one of the following metals does not react with water or steam?
   A   Aluminium.
   B   Iron.
   C   Magnesium.
   D   Silver.

8. Which one of the following non-metals reacts with water to produce oxygen?
   A   Chlorine.
   B   Carbon.
   C   Phosphorus.
   D   Sulphur.

Chapter 11

Acids and bases

Acids and bases have already been introduced in Chapter 5, but here we shall consider their properties in more detail.

Ionisation of acids

The common acids encountered in the laboratory are nitric acid, HNO_3, sulphuric acid, H_2SO_4, ethanoic (or acetic) acid, CH_3COOH, and hydrochloric acid, which is an aqueous solution of hydrogen chloride gas, HCl. In the absence of water, these (and all other) acids are covalent compounds which contain one or more hydrogen atoms in their molecules. When they are dissolved in water, however, their molecules split up to produce protons and negative ions. An example is shown in Fig. 11.1. This formation of ions from molecules is known as *ionisation*.

The protons and negative ions in solution have water molecules attached to them and are said to be *hydrated*. This is signified by writing (aq) after the ion. The H^+(aq) ion is called the *hydrogen ion*.

$HCl = H^+(aq) + Cl^-(aq)$
$HNO_3 = H^+(aq) + NO_3^-(aq)$
$H_2SO_4 = 2H^+(aq) + SO_4^{2-}(aq)$
$CH_3COOH = H^+(aq) + CH_3COO^-(aq)$

It is the presence of hydrogen ions in solution which gives acids their characteristic properties. An acid can therefore be defined as a substance which ionises in aqueous solution to form hydrogen ions.

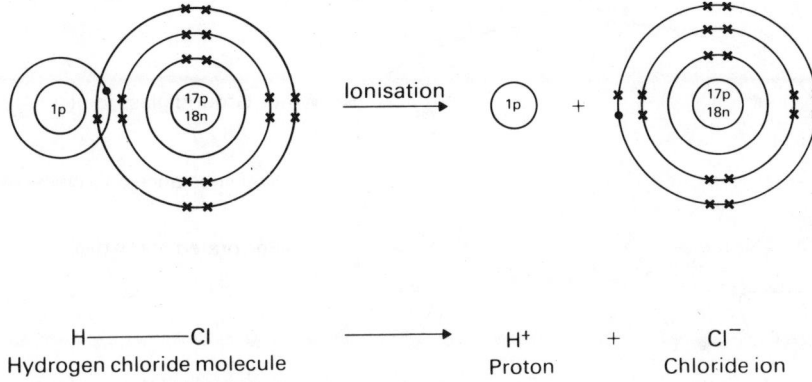

| H————Cl | ⟶ | H⁺ | + | Cl⁻ |
Hydrogen chloride molecule ⟶ Proton + Chloride ion

Fig. 11.1 Ionisation of the hydrogen chloride molecule

Strong and weak acids

The extent of ionisation varies greatly from one acid to another. Some acids, such as HNO_3, H_2SO_4 and HCl, ionise to a large extent and thus produce a high concentration of hydrogen ions in aqueous solution. They are called *strong acids*. Others, such as CH_3COOH, ionise to only a small extent and give only a low concentration of hydrogen ions in aqueous solution. They are termed *weak acids*.

Basicity of acids

It will be seen from the equations printed above that when sulphuric acid ionises two hydrogen ions are formed from one molecule, but when ethanoic acid, CH_3COOH, ionises only one hydrogen ion per molecule is produced despite the fact that each molecule of the acid contains four hydrogen atoms.

Acids, such as HCl, HNO_3 and CH_3COOH, which yield one hydrogen ion per molecule are termed *monoprotic* or *monobasic*, and those, *e.g.* H_2SO_4, which produce two hydrogen ions per molecule are described as *diprotic* or *dibasic*. The number of hydrogen ions which are formed from one molecule is known as the *basicity* of the acid.

Mineral acids

The common inorganic or *mineral acids*, *i.e.* nitric acid, hydrochloric acid and sulphuric acid, are extremely corrosive, especially in concentrated form. They should therefore be handled carefully, and any splashes or spillages immediately washed with plenty of water.

Sulphuric acid

Sulphuric acid is used in the laboratory in both concentrated and dilute forms. *Concentrated sulphuric acid* contains 98 per cent of the acid and 2 per cent of water. It is a colourless, syrupy liquid with a great attraction for water. *Dilute sulphuric acid* contains about 2 mol dm⁻³ of the acid.

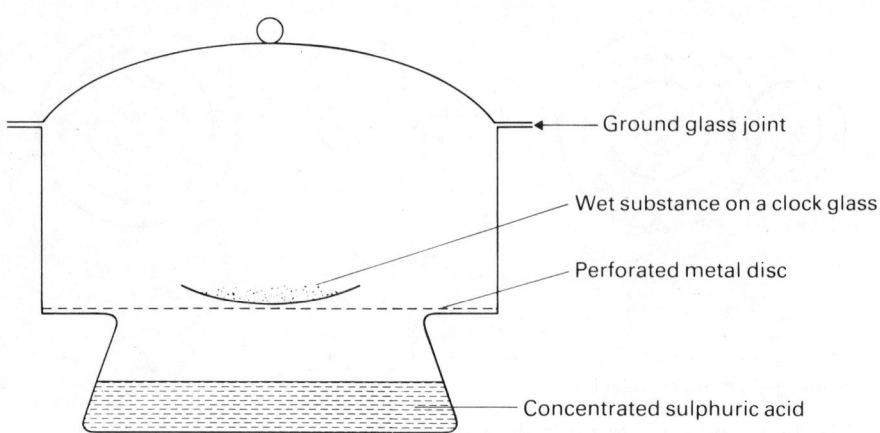

Fig. 11.2 Drying a substance in a desiccator

Care must be taken when diluting concentrated sulphuric acid because of the large amount of heat that is given out. **The acid should always be added to water** while the latter is continuously stirred and cooled. Both the acid and the water should be cold before they are mixed together.

Because of its attraction for water, concentrated sulphuric acid is used as a drying agent and a dehydrating agent. It is said to be a *drying agent* because it removes water from wet substances when it comes into contact with them. To avoid contaminating the wet substance with the sulphuric acid we use an apparatus called a *desiccator* (see Fig. 11.2). Water evaporates from the wet substance, passes through the perforated platform of the desiccator, and is absorbed by concentrated sulphuric acid in the base.

Gases, such as oxygen, nitrogen, hydrogen, carbon dioxide, hydrogen chloride, chlorine and carbon monoxide, can be dried by simply bubbling them through concentrated sulphuric acid.

Concentrated sulphuric acid is also described as a *dehydrating agent* because it can absorb water from substances which are not necessarily wet. It does this by breaking chemical bonds in the substances and removing hydrogen and oxygen atoms in a 2:1 ratio. Human skin is attacked in this way by concentrated sulphuric acid, which is why splashes of the acid must be quickly washed off with plenty of water.

The dehydrating properties of concentrated sulphuric acid are utilised in the laboratory for preparing various substances. Ethene, for example, can be prepared by the dehydration of ethanol. Concentrated sulphuric acid removes a molecule of water from a molecule of ethanol, C_2H_5OH, to give a molecule of ethene, C_2H_4:

$$C_2H_5OH - H_2O = C_2H_4$$

Similarly charcoal, a pure form of carbon, can be prepared by dehydrating a sugar such as sucrose, $C_{12}H_{22}O_{11}$:

$$C_{12}H_{22}O_{11} - 11H_2O = 12C$$

Concentrated sulphuric acid also possesses oxidising properties, *i.e.* it has a tendency to oxidise other substances. This aspect is discussed in the next chapter.

Nitric acid

Nitric acid, like sulphuric acid, is used in the laboratory in both concentrated and dilute forms. Concentrated nitric acid is an approximately 70 per cent solution of the acid in water, while dilute nitric acid usually contains about 4 mol dm^{-3} of HNO_3. Solutions of nitric acid are colourless when fresh, but slowly turn yellow when exposed to sunlight owing to decomposition.

Concentrated nitric acid possesses strong oxidising properties (see Chapter 12). Care should thus be taken not to mix concentrated nitric acid with substances which are easily oxidised, because violent reactions may occur.

Hydrochloric acid

Concentrated hydrochloric acid is a saturated solution of hydrogen chloride in water. It contains approximately 36 per cent of HCl. Unlike concentrated nitric acid or concentrated sulphuric acid, this acid possesses only weak oxidising properties and has no dehydrating properties at all. The dilute acid, approximately 4 mol dm^{-3} in concentration, is often used in the laboratory.

Bases

Bases have already been introduced in Chapter 5, but here their properties will be discussed in more detail.

A base is essentially a substance which will accept and combine with hydrogen ions from an acid. Thus, when a base reacts with an acid in solution, hydrogen ions are removed from the solution and its acidic character is lost. Such a reaction between an acid and a base is described as *neutralisation*.

The commonest bases are the electrovalent oxides and hydroxides of metals, and also ammonia. Metal oxides contain the oxide ion, O^{2-}, while the hydroxides contain the hydroxide ion, OH^-. Ammonia, in contrast, is covalent.

Insoluble bases

Most bases are insoluble in water. Examples are magnesium oxide, MgO, magnesium hydroxide, $Mg(OH)_2$, and copper(II) oxide, CuO. In neutralisation reactions involving these compounds, oxide or hydroxide ions from the base combine with hydrogen ions from the acid to give molecules of water, thus:

$$O^{2-} + 2H^+(aq) = H_2O$$
$$OH^- + H^+(aq) = H_2O$$

Soluble bases

Water-soluble bases, called *alkalis*, include the well known compounds sodium hydroxide, NaOH, potassium hydroxide, KOH, and ammonia, NH_3. They can be defined as compounds which give rise to hydroxide ions in aqueous solution (*cf.* acids, which give rise to hydrogen ions in aqueous solution). There are two clearly defined structural types, as follows.

i.   Electrovalent bases, which give rise to hydroxide ions in solution by *dissociation, i.e.* the separation of ions (see p. 77). For example,

$$NaOH(s) = Na^+ (aq) + OH^- (aq)$$

We write OH^- (aq) rather than just OH^- because in solution the hydroxide ion (like the hydrogen ion) has water molecules attached to it.

ii.   Covalent bases, which give rise to hydroxide ions in solution by *ionisation, i.e.* by chemical reaction with the water. Ammonia falls into this category. When ammonia dissolves in water a small proportion of ammonia molecules reacts with the water to form ammonium ions and hydroxide ions:

$$NH_3 + H_2O = NH_4^+(aq) + OH^-(aq)$$

Soluble bases, like acids, are classified as 'strong' and 'weak'. *Strong bases* are those which give rise to a high concentration of hydroxide ions in aqueous solution. They include sodium hydroxide and potassium hydroxide, which are completely dissociated in solution. *Weak bases,* in contrast, give a low concentration of hydroxide ions in aqueous solution. Ammonia is a weak base because it ionises in water to only a very small extent.

The neutralisation of an acid by a strong base can be represented by the equation:

$$H^+(aq) + OH^-(aq) = H_2O(l)$$

However, the corresponding reactions with weak bases are rather different. For example, in a solution of ammonia there are far fewer hydroxide ions than ammonia molecules; consequently, the principal reaction to occur when ammonia neutralises an acid is as follows:

$$NH_3 + H^+(aq) = NH_4^+(aq)$$

The safe use of alkalis

The solid hydroxides of sodium or potassium and their concentrated aqueous solutions are said to be 'caustic', because they rapidly attack the skin and all other animal and vegetable tissues. Alkalis should therefore never be allowed to come into contact with such materials. Sodium hydroxide pellets, for example, must never be handled directly. If, by accident, alkalis do come into contact with the skin, the affected area should first be washed with plenty of water and afterwards treated with a weak acid, such as dilute ethanoic acid, to neutralise any remaining hydroxide.

It is extremely important to avoid splashes of alkali in the eye, because the cornea, *i.e.* the front covering of the eye, can become opaque, leading to blindness. **In any such accident, immediately after first aid treatment, medical attention must be sought.**

Concentrated solutions of ammonia are particularly dangerous when they come into contact with the eye. Safety goggles should always be worn when handling concentrated solutions of this compound.

The pH scale

All aqueous solutions contain hydrogen ions and hydroxide ions, no matter what else is present in solution.

In a *neutral solution* the hydrogen ion concentration equals the hydroxide ion concentration, *i.e.*

$$[H^+(aq)] = [OH^-(aq)]$$
In short, $[H^+] = [OH^-]$

Square brackets denote concentration in mol dm^{-3}, *i.e.* molarity.

In an *acidic solution* $[H^+]$ is greater than $[OH^-]$. If a solution is weakly acidic $[H^+]$ may be only slightly greater than $[OH^-]$, but in a strongly acidic solution $[H^+]$ is very much greater than $[OH^-]$. Nevertheless, no matter how strongly acidic a solution may be, it still contains some hydroxide ions.

In a *basic* or *alkaline solution* $[OH^-]$ is greater than $[H^+]$, although in a weakly basic solution $[OH^-]$ may be only slightly greater than $[H^+]$. In a strongly basic solution $[OH^-]$ is very much greater than $[H^+]$, but no matter how strongly basic a solution may be it still has a small but measurable concentration of hydrogen ions.

Hydrogen ion concentration therefore provides a measure of both acidity and alkalinity, for relatively high values tell us that a solution is acidic while low values denote alkalinity. Hydroxide ion concentration can be used in a corresponding manner.

Instead of quoting the hydrogen ion concentration of a solution in moles per cubic decimetre, it is customary to represent acidity or alkalinity by a *pH value*. This is a number (without units) related to $[H^+]$ *in an inverse manner*, *i.e.* the greater the $[H^+]$ of a solution, the lower is its pH.

A neutral solution has a pH of 7. Acidic solutions have pH values ranging between 0 and 7, whereas for basic solutions the pH lies between 7 and 14. The pH scale is thus a series of numbers extending from 0 to 14.

increasing $[H^+]$ and decreasing $[OH^-]$

decreasing $[H^+]$ and increasing $[OH^-]$

| pH | 0 | 1 | 2 | 3 | 4 | 5 | 6 | 7 | 8 | 9 | 10 | 11 | 12 | 13 | 14 |

increasingly acidic solutions — NEUTRAL — increasingly basic solutions

STRONG ACIDS   WEAK ACIDS   WEAK BASES   STRONG BASES

Solutions of strong acids, such as hydrochloric acid, nitric acid and sulphuric acid, always have low pHs. For example, hydrochloric acid solutions of concentration 1 mol dm^{-3} and 0.1 mol dm^{-3} have pH values of 0 and 1 respectively. Weak acids never give solutions of such low pH; *e.g.* for ethanoic acid of concentration 0.1 mol dm^{-3} the pH is approximately 3.

Strong bases, such as sodium hydroxide and potassium hydroxide, give solutions of very high pH. Solutions of concentration 1 mol dm^{-3} and 0.1 mol dm^{-3} have pHs of 14 and 13 respectively. Weak bases in solution usually have pH values lying between 10 and 12; *e.g.* aqueous ammonia of concentration 0.1 mol dm^{-3} has a pH of approximately 11.

Determination of pH

The approximate pH of a solution can be determined by means of *pH papers*. These are strips of absorbent paper, impregnated with a mixture of indicators (see p. 150) known as *universal indicator*. They show a different colour at each pH as illustrated by the colour chart which is supplied with them.

pH papers are of two kinds, namely 'full range' and 'narrow range'. Full range papers cover most of the scale, from pH 1 to pH 14, whereas narrow range papers cover only a small part of it, *e.g.* from pH 6 to pH 8.

Full range papers are relatively inaccurate, but are always used first to obtain a rough estimate of pH. A strip of the paper is dipped into the solution whose pH is required, and the colour of the paper is compared with the colours on the chart. The appropriate narrow range paper is then selected and the procedure is repeated.

Although pH papers provide only an approximate measure of pH, they are cheap and reliable and have the outstanding merit of being simple and quick to use.

Summary

At the conclusion of this chapter, you should be able to:
1. describe an acid as a substance which yields hydrogen ions in solution,
2. distinguish between strong and weak acids,
3. state the basicity of common acids,
4. recognise the corrosive nature of acids, and the hazards associated with their use,
5. recognise the special dangers associated with concentrated sulphuric acid, because of its affinity for water, and nitric acid, because of its oxidising powers,
6. recognise that concentrated sulphuric acid is a drying agent and a dehydrating agent,
7. describe a base as a substance which accepts hydrogen ions from an acid and thus removes its acidic properties,
8. describe an alkali as a water-soluble base which yields hydroxide ions in solution,
9. state that ammonia is a base,
10. describe neutralisation in terms of

$$H^+(aq) + OH^-(aq) = H_2O(l),$$

11. describe the hazards associated with alkalis and their solutions and with concentrated aqueous ammonia,
12. describe pH as a numerical scale and recognise what values relate to acidic solutions, basic solutions and neutral solutions,
13. use pH papers to determine the approximate pH of a solution.

Questions

1. State whether each of the following acids and bases is strong or weak, and give the names and formulae of the ions produced on ionisation.
   (a) Hydrogen chloride   (d) Ethanoic acid
   (b) Sulphuric acid   (e) Sodium hydroxide
   (c) Nitric acid   (f) Ammonia

   In questions 2–7, select the most appropriate answer, labelled A, B, C or D.

2. Which one of the following acids is dibasic?
   A   Hydrochloric acid.
   B   Nitric acid.
   C   Sulphuric acid.
   D   Ethanoic acid.

3. For the dehydration of ethanol to ethene, a suitable dehydrating agent is concentrated:
   A   hydrochloric acid,
   B   sulphuric acid,
   C   nitric acid,
   D   ethanoic acid.

4. Which one of the following equations represents the principal reaction that occurs between ammonia and hydrochloric acid?
   A $OH^- (aq) + H^+ (aq) = H_2O$
   B $O^{2-} + 2H^+(aq) = H_2O$
   C $NH_4OH + HCl = NH_4Cl + H_2O$
   D $NH_3 + H^+(aq) = NH_4{}^+$

5. In a neutral aqueous solution,
   A $[H^+]$ is equal to $[OH^-]$,
   B $[H^+]$ is greater than $[OH^-]$,
   C $[H^+]$ is smaller than $[OH^-]$,
   D   there are no H^+ or OH^- ions present.

6. A solution of ethanoic acid of concentration 0.1 mol dm^{-3} would be expected to have a pH of approximately:
   A   1,
   B   3,
   C   7,
   D   11.

7. In which one of the following series are the compounds arranged in increasing order of pH value? (Assume that each compound is in solution of concentration 0.1 mol dm^{-3}.)
   A   Ammonia, ethanoic acid, hydrochloric acid, sodium hydroxide.
   B   Ethanoic acid, ammonia, hydrochloric acid, sodium hydroxide.
   C   Ammonia, ethanoic acid, sodium hydroxide, hydrochloric acid.
   D   Hydrochloric acid, ethanoic acid, ammonia, sodium hydroxide.

8. Which of the following statements are true and which are false?
   (a) The pH of an acidic solution is greater than 7.
   (b) No solution can have a pH of 0.
   (c) Ammonia solutions have a pH of less than 7.

(d) Alkaline solutions have a pH of greater than 7.

(e) A concentrated solution of ammonia is said to be a strong base.

(f) A dilute solution of hydrochloric acid is said to be a weak acid.

(g) As the hydrogen ion concentration of a solution rises, its hydroxide ion concentration falls.

(h) The correct first aid treatment for concentrated sulphuric acid burns is plenty of water.

Chapter 12

Types of chemical reaction

The principal types of chemical reaction are as follows:
>   Neutralisation
>   Double decomposition
>   Oxidation-reduction
>   Decomposition
We shall discuss each in turn.

Neutralisation

The idea of neutralisation has been introduced in earlier chapters. We have seen that it is essentially the reaction between an acid and a base to produce a salt. Water is usually formed at the same time.

There is a problem, when carrying out such reactions in the laboratory, in that we cannot see when the acid has been neutralised by the base. This is resolved by a *titration*, in which we determine, with the help of a coloured substance called an *indicator*, the exact volume of one solution which is required to react completely with a given volume of another solution.

Titrations may involve various types of reaction, but in this book we are concerned only with *acid-base titrations*, which all depend on neutralisation.

Acid-base titrations

To perform an acid-base titration, we take an accurately measured volume of one solution (*e.g.* the acid), add a few drops of indicator, and gradually run in the other solution (the base, in this example) from a graduated tube called a 'burette'. At the *end-point* of the titration, *i.e.* when the acid has just been neutralised by the base, the indicator changes colour. From the graduations on the burette we can then tell the exact volume of base solution that is required.

Indicators

An acid-base indicator is essentially a compound whose colour is sensitive to changes in pH. In other words, it shows one colour in acidic solution (low pH), another colour in basic solution (high pH), and an intermediate colour in neutral solution.

The best known indicator is litmus, which is red in acidic solution, blue in basic solution and purple in neutral solution. Litmus, however, is unsuitable for use in acid-base titrations because its colour change at the end-point is insufficiently sharp.

The most satisfactory indicators show a marked colour contrast between the two extremes, and change colour very abruptly at the end-point. Three of the commonest used in titrations are shown in Table 12.1.

Table 12.1 Acid-base indicators

Indicator	Colour in acidic solution	Colour in basic solution	Colour in neutral solution
Phenolphthalein	Colourless	Red	Pale pink
Methyl orange	Red	Yellow	Orange
Methyl red	Red	Yellow	Orange

Apparatus for measuring volumes of liquids

Measuring cylinders (see Fig. 12.1) are used only for measuring approximate volumes. They are available in various sizes (10 cm³, 25 cm³, 50 cm³, 100 cm³, etc.), and it is always advisable to select the smallest that is appropriate for a particular purpose. For example, 7 cm³ of liquid can be measured more accurately in a 10 cm³ measuring cylinder than in a 100 cm³ one, and the former is therefore preferable.

To measure the volume of a liquid to the sufficiently high degree of accuracy required for titrations we use pipettes and burettes (see Fig. 12.1). A *pipette* is essentially a glass tube, with a jet at one end and a bulb in the middle, which is calibrated to deliver a specified volume of liquid at a given temperature. Pipettes, like measuring cylinders, are available in a variety of sizes, but the commonest delivers *exactly* 25 cm³ of liquid at room temperature.

A pipette is filled by first attaching a *pipette filler* to the upper end. Various types of pipette filler are available, but the commonest consists of a rubber bulb equipped with valves. By squeezing the valves, the flow of air into and out of the bulb can be controlled. Liquid is drawn into the pipette by squeezing the bulb, dipping the jet of the pipette in the liquid, and then releasing one of the valves. **To avoid getting liquid in the mouth, you should never fill a pipette by sucking at the upper end.**

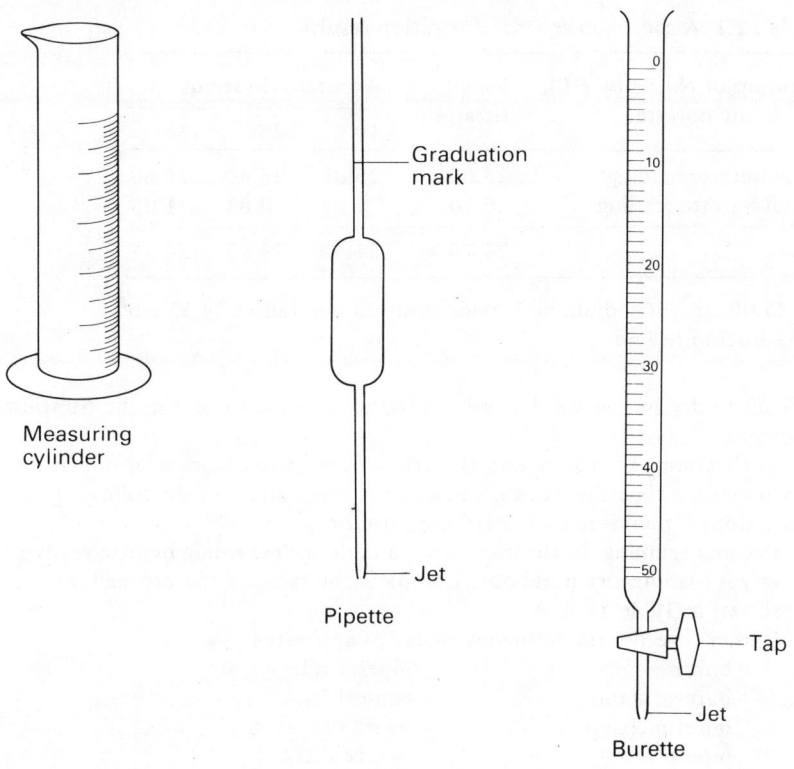

Fig. 12.1 Apparatus for measuring volumes of liquids

A *burette* consists of a glass tube, open at the top but with a tap and a jet at the bottom. The tube is calibrated in tenths of a cubic centimetre, from 0 cm^3 at the top to 50 cm^3 at the bottom. Burettes, unlike pipettes, are filled simply by pouring liquid from a beaker into the upper end. A small filter funnel can be placed in the top of the burette to avoid spillage, but if so the funnel must be removed before the burette is used.

When any graduated glassware is used, it is important to remember the following things.

   i. The correct volume is given by the graduation mark coinciding with the *bottom* of the liquid meniscus.

  ii. Whenever readings are made (or whenever the volume in a pipette is checked) the apparatus should be held so that the liquid meniscus is at eye level.

 iii. Apparatus must never be heated or cooled.

Instructions for performing a titration

Throughout the following instructions it is assumed that all glassware is chemically clean. If there is any doubt, the apparatus should be cleaned with detergent and then thoroughly rinsed with water. Pipettes and burettes are

Table 12.2 A specimen record of titration results

Titration of NaOH by HCl. HCl in the burette	Rough titration	Accurate titrations		
		1st	2nd	3rd
Final burette reading	25.00	25.85	25.40	25.60
Initial burette reading	0.30	1.25	0.85	1.05
Titre	24.70	24.60	24.55	24.55

i.e. 25.00 cm^3 of sodium hydroxide solution neutralises 24.55 cm^3 of hydrochloric acid

difficult to dry but, as we shall see, allowance is made for this in the titration procedure.

In titrations, as in all practical work, a permanent record of results must be made at the time the work is carried out. Paragraph 1 of the following instructions is therefore particularly important.

1. Before beginning the titration, draw a table for recording burette readings in your laboratory notebook. Complete the table as you proceed, as shown in Table 12.2.
2. Gather together the following pieces of apparatus:

   burette   pipette filler
   burette stand   conical flask
   burette clamp   white tile
   pipette   two beakers
3. Collect about 100 cm^3 of both acid and base in the beakers, which must be labelled. You will find on most beakers a white spot on which you can pencil the contents. Failing this, use a felt-tipped glass marker or a self-adhesive label.
4. Decide which solution is to go in the pipette, and which in the burette, and record your decision in your notebook.

   If the burette has a PTFE plastic tap it does not matter which solution it contains. Glass taps, however, can become jammed by alkali, and if the burette is fitted with one of these it is generally advisable for it to hold acid.

   A basic solution must never be allowed to remain in a burette for a long period of time, because this will etch the glass.
5. Prepare the burette by washing it out with the solution which is to go in it. (This is to remove residual water.) Pour in the solution so that the burette is about half full, then rotate it in an almost horizontal position so that the solution comes into contact with as much of the inside surface as possible. Drain the burette through the jet and repeat the process.
6. Clamp the burette in a burette stand and fill it ready for use to just above the 0 cm^3 mark. Place the beaker containing the same solution under the burette, and open the tap fully to fill the jet with liquid. Make sure that no air remains trapped in the jet. (In particularly stubborn cases, air

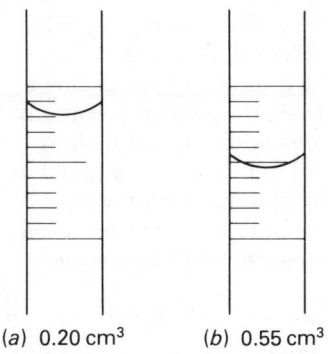

(a) 0.20 cm³   (b) 0.55 cm³

Fig. 12.2 Correct reading of a burette

bubbles can be dislodged by shaking the burette carefully as the solution runs out.)

7. Record the initial burette reading to the nearest 0.05 cm³. For this purpose you may have to lift the burette down to your level. Hold a piece of white paper behind the burette as you take the reading: this makes it easier to see the bottom of the meniscus.

   You cannot read a burette accurately to 0.05 cm³, but you will find that the liquid meniscus lies either on a division (see Fig. 12.2(a)) or between two divisions (see Fig. 12.2(b)). In the former case the second decimal place is 0, and in the latter you should estimate it as 5.

   It is bad laboratory practice to waste time ensuring that the initial reading is exactly zero.

8. Prepare the conical flask for use by rinsing it out, first with tap water and then with distilled water. Residual distilled water does not matter, for it will make no difference to the end-point of the titration.

9. Prepare the pipette in the same way as the burette, by washing it out with the solution which is to go in it. Half fill the pipette with a pipette filler, and rotate it horizontally so that the solution comes into contact with all the inside surface to about 3 cm above the calibration mark. Drain the pipette through the jet and repeat the procedure.

10. Fill the pipette, again with a pipette filler, until its level is above the calibration mark. Remove the filler and quickly place your finger over the top to stop the solution running out. Hold the pipette so that the jet is above the level of the solution in the beaker, and gradually release your finger so that the liquid level slowly falls until the bottom of the meniscus coincides with the calibration mark. Immediately increase your finger pressure so that the liquid level remains exactly at this mark.

   Should the level fall too far, you must draw some more solution into the pipette and try again. If you experience difficulty in controlling the pipette, it may be that your finger is wet. Both your finger and the top of the pipette should be reasonably dry.

If preferred, the adjustment of level in a pipette, and the draining (see below), can be carried out with the pipette filler still in place.

11. Allow the pipette to drain by gravity into the conical flask, until only a small amount of solution remains in the jet. Drain the jet by momentarily dipping it just below the surface of the solution in the conical flask. A drop of solution will still be visible in the jet, but it should be allowed to stay there. **You must never drain a pipette by blowing down the top.** Allowance is made for the last drop of liquid in the jet when a pipette is calibrated, and when used as instructed the apparatus will deliver the exact volume that is printed on its bulb.

    Until it is needed again, put the pipette at the back of the bench or rest it on a tripod so that it will not roll on to the floor.

12. Add two or three drops of indicator, such as methyl orange or phenol-phthalein, to the solution in the conical flask. Avoid using too little or too much indicator, because the change in colour at the end-point of the titration is difficult to observe if the colours are very faint or very intense. The solution in the flask is now ready to be titrated with the other solution from the burette.

13. Place the conical flask underneath the burette, and adjust the height of the burette so that the jet is just above the rim of the flask. Allow the conical flask to stand on the white tile so that the colour of the solution can be clearly seen.

    Begin with a rough titration, *i.e.* one carried out fairly rapidly to provide an approximate end-point. Open the tap and allow the *titrant, i.e.* the solution in the burette, to run into the conical flask. Swirl the latter continuously, to make sure that the two liquids become properly mixed together. Near the end-point, when the solution in the flask begins to show marked signs of a colour change, reduce the rate at which the titrant is added. Watch for the colour to change permanently, and as soon as this happens stop the titration and note the final burette reading. Subtraction of the initial burette reading from the final reading gives the *titre, i.e.* the volume of solution delivered from the burette. The approximate end-point will almost always be an over-titration.

    Discard the solution and wash out the conical flask.

14. Refill the burette and perform another titration with a fresh portion of solution in the conical flask. The aim in this case is to obtain an accurate end-point. Titrate rapidly to about 1 cm^3 below the approximate end-point, and then run in the titrant very gradually. Swirl well all the time. In the immediate vicinity of the end-point, when the solution in the conical flask is slow to revert to its original colour after the burette tap has been closed, add the titrant dropwise and try to avoid over-titrating by even one drop.

    Every single drop of titrant matters. The addition of one drop of solution increases the burette reading by about 0.05 cm^3. Thus, if you over-titrate by two drops, and obtain a titre of, say, 25.10 cm^3 instead of

25.00 cm^3, the error is $\frac{0.1}{25} \times 100 = 0.4$ per cent, which is the limit of

what is acceptable.

If you are doubtful whether you have reached the end-point there are two things you can do.

Read the burette and then add one more drop of titrant. If this results in over-titration, the original burette reading was correct. If, however, the additional titrant produces no obvious colour change, the end-point has not yet been reached.

ii. Take two test tubes, one containing acid and the other base, and stand them in a test tube rack. Put one or two drops of indicator in each, so that you can see the two colour extremes. Aim at an intermediate colour. (If phenolphthalein is the indicator, and you are running base into acid, you should aim at a faint pink colour that persists for about 30 seconds after the flask has been swirled.)

15. Carry out further accurate titrations until you get at least two results which agree with one another to within 0.10 cm^3. Write a summary of your accurate findings underneath the table of results, as shown in Table 12.2.

Assessment of results

Wherever possible, try to avoid taking an average of titration results. If you do so, you will be averaging results which may well be right with others which are probably over-titrations, and you are likely to get an inaccurate figure. Instead of this, you should continue titrating until you get concordant results on which you can rely.

You are justified in averaging results only if you run short of time or material. Even then, 'rogue' results, which are obviously wrong, should be ignored. Suppose, for instance, that the following set of 'accurate' titres has been obtained and that no further experimental work is possible:

27.40 cm^3 27.80 cm^3 27.60 cm^3 27.45 cm^3 27.55 cm^3

27.80 cm^3 is probably an over-titration and can be neglected. The other figures can be averaged thus:

$$\frac{27.40 + 27.60 + 27.45 + 27.55}{4} = 27.50 \text{ cm}^3$$

However, the treatment would have been different had the results been as follows:

27.40 cm^3 27.80 cm^3 27.60 cm^3 27.45 cm^3 27.45 cm^3

In this case it can be argued that both 27.80 cm^3 and 27.60 cm^3 represent over-titrations, so that the true end-point is 27.45 cm^3.

Volumetric analysis

A titration is usually carried out to find the concentration of one of the solutions involved. If the concentration of one solution (*e.g.* the base) is known, that of the other (*e.g.* the acid) can readily be calculated from the titration results. Work of this kind is known as *volumetric analysis*.

By way of illustration, let us use the specimen results shown in Table 12.2, and suppose that the concentration of the sodium hydroxide is 0.099 mol dm^{-3} (0.099 M), while that of the hydrochloric acid is unknown. By looking at the chemical equation for the reaction it is not difficult to work out the concentration (molarity) of the hydrochloric acid.

$$NaOH + HCl = NaCl + H_2O$$

The equation tells us that sodium hydroxide and hydrochloric acid react together in a 1:1 molar ratio, *i.e.*

$$\frac{\text{moles of NaOH}}{\text{moles of HCl}} = \frac{1}{1} \tag{1}$$

The number of moles of solute in a solution, *i.e.* the amount of dissolved substance, depends partly on the concentration (molarity) of the solution and partly on its volume.

If a certain solution has a concentration (molarity) of n mol dm^{-3}, then 1 dm^3 of solution contains n mol of solute, and V dm^3 of solution contain $n \times V$ mol of solute.

To summarise, **the number of moles of solute in a solution is given by the concentration (molarity) multiplied by the volume in cubic decimetres.** If the volume is in cm^3 rather than dm^3, we must modify this as follows:

$$\text{number of moles} = \text{concentration (molarity)} \times \frac{\text{volume in cm}^3}{1000}$$

Substitution in equation [1] gives:

$$\frac{c_{NaOH} \times \dfrac{V_{NaOH}}{1000}}{c_{HCl} \times \dfrac{V_{HCl}}{1000}} = \frac{1}{1}$$

where c = concentration in mol dm^{-3} (molarity) and V = volume in cm^3. The thousands cancel to give:

$$\frac{c_{NaOH} \times V_{NaOH}}{c_{HCl} \times V_{HCl}} = \frac{1}{1} \tag{2}$$

In our example, 25 cm^3 of sodium hydroxide solution of concentration 0.099 mol dm^{-3} (0.099 M) neutralises 24.55 cm^3 of hydrochloric acid of unknown concentration (molarity). Substitution of these figures in equation [2] gives:

$$\frac{0.099 \times 25}{c_{HCl} \times 24.55} = \frac{1}{1}$$

therefore $c_{HCl} = \dfrac{0.099 \times 25 \times 1}{24.55 \times 1} = 0.101$ mol dm^{-3} (0.101 M)

Expressing the result as a mass concentration is easy, once we remember that to convert moles to grams we multiply by relative molecular mass. The relative molecular mass of hydrogen chloride is 36.5, therefore mass concentration of the hydrochloric acid = 0.101×36.5 g dm^{-3} = 3.69 g dm^{-3}.

Sodium carbonate is often used as the base in acid-base titrations, because the anhydrous solid is suitable as a *primary standard*, *i.e.* a reliable reference compound in volumetric analysis. Anhydrous sodium carbonate is readily available in a high state of purity, and can be weighed out on an analytical balance and then dissolved in distilled water to make a standard solution, *i.e.* one whose concentration is accurately known (see p. 82). When titrated against hydrochloric acid with methyl orange as the indicator, the following reaction occurs:

$Na_2CO_3 + 2HCl = 2NaCl + H_2O + CO_2$

The equation tells us that sodium carbonate and hydrochloric acid react together in a 1:2 molar ratio, *i.e.*

$$\dfrac{\text{moles Na}_2\text{CO}_3}{\text{moles HCl}} = \dfrac{1}{2}$$

therefore $\dfrac{c_{Na_2CO_3}}{c_{HCl}} \times \dfrac{V_{Na_2CO_3}}{V_{HCl}} = \dfrac{1}{2}$

Hence, if the titration gives us the volume of each solution, and a standard solution of sodium carbonate has been used, the concentration (molarity) of the acid can be calculated.

Double decomposition

When two compounds, often in solution, react together to form two new compounds by interchanging ions, the type of reaction is called *double decomposition*. The general equation for such a reaction is:

$AB + CD = AD + CB$

The reactants AB and CD must furnish ions in solution. A and C represent positively charged ions; these may be metal ions, ammonium ions or hydrogen ions. B and D represent negative ions. Essentially all that happens in a double decomposition reaction is that the positive ions from each reactant combine with the negative ions of the other.

At least one of the reactants must be a salt. The other reactant can be another salt, or it can be an acid or a base. This gives us three possible types of double decomposition reaction.

(i)   **acid** + **salt** = **another acid** + **another salt**
   e.g. $2HCl + FeS = H_2S$ $+ FeCl_2$

(ii)　　　**base　＋ salt**　　　**= another base ＋ another salt**
　　e.g. $2NaOH + Pb(NO_3)_2 = Pb(OH)_2 + 2NaNO_3$
(iii)　　　**salt　＋ salt**　　　**= two other salts**
　　e.g. $NaCl + AgNO_3 = AgCl + NaNO_3$

Double decomposition reactions are not always useful as a method of preparing salts because they may not go to completion, *i.e.* in some cases the reactants are not completely converted into products. For example, if solutions of sodium chloride and potassium nitrate are mixed together, the resulting solution contains sodium chloride, potassium nitrate, potassium chloride and sodium nitrate:

$$NaCl + KNO_3 = KCl + NaNO_3$$

Consequently, this does not provide us with a useful way of making potassium chloride or sodium nitrate.

If a double decomposition is to be useful, two ions from the reactants must combine together in such a way that these ions are permanently removed from solution. The ions are then not available to take part in the reverse change, and the double decomposition is able to proceed to completion from left to right.

In practice, pairs of oppositely charged ions can be removed from solution by forming:
i.　insoluble compounds (salts or bases),
ii.　unstable compounds,
iii.　volatile compounds (acids or covalent bases),
iv.　weak acids.
We shall now consider each situation in turn.

The formation of insoluble compounds (salts or bases)

Let us consider the reaction which occurs in aqueous solution between lead(II) nitrate and sodium chloride. A white *precipitate* of lead(II) chloride is formed, which can be filtered off, washed and dried. (The term 'precipitate' is applied to any insoluble compound formed by chemical reaction in solution.) The filtrate contains sodium nitrate, a soluble salt which can be obtained from solution by crystallisation.

$$\underbrace{Pb^{2+} + 2NO_3^-}_{\substack{\text{lead(II) nitrate}\\\text{solution}}} + \underbrace{2Na^+ + 2Cl^-}_{\substack{\text{sodium chloride}\\\text{solution}}} = \underbrace{PbCl_2(s)}_{\substack{\text{insoluble}\\\text{lead(II) chloride}}} + \underbrace{2Na^+ + 2NO_3^-}_{\substack{\text{sodium nitrate}\\\text{solution}}}$$

The essential change involves the combination of lead(II) ions and chloride ions to give solid lead(II) chloride, and is best represented by writing an ionic equation:

$$Pb^{2+}(aq) + 2Cl^-(aq) = PbCl_2(s)$$

Sodium ions and nitrate ions do not feature in the ionic equation as they play no direct part in the reaction. They are termed 'spectator ions' (see p. 110).

Similarly, in the reaction between barium chloride and sulphuric acid, barium ions combine with sulphate ions to form insoluble barium sulphate:

$$\underbrace{Ba^{2+} + 2Cl^-}_{\substack{\text{barium chloride} \\ \text{solution}}} + \underbrace{2H^+ + SO_4^{2-}}_{\substack{\text{sulphuric acid} \\ \text{solution}}} = \underbrace{BaSO_4(s)}_{\substack{\text{insoluble} \\ \text{barium sulphate}}} + \underbrace{2H^+ + 2Cl^-}_{\substack{\text{hydrochloric acid} \\ \text{solution}}}$$

Cancellation of spectator ions gives the ionic equation:

$$Ba^{2+}(aq) + SO_4^{2-}(aq) = BaSO_4(s)$$

The formation of unstable compounds

The reaction between carbonates and acids is typical of this type of double decomposition. For example, when sodium carbonate reacts with sulphuric acid, unstable carbonic acid is formed:

$$\underbrace{2Na^+ + CO_3^{2-}}_{\text{sodium carbonate}} + \underbrace{2H^+ + SO_4^{2-}}_{\text{sulphuric acid}} = \underbrace{2Na^+ + SO_4^{2-}}_{\text{sodium sulphate}} + \underbrace{H_2CO_3}_{\text{carbonic acid}}$$

Carbonic acid decomposes almost immediately into water and carbon dioxide gas:

$$H_2CO_3 = H_2O + CO_2$$

The overall reaction is thus:

$$Na_2CO_3 + H_2SO_4 = Na_2SO_4 + H_2O + CO_2$$

In ionic terms,

$$CO_3^{2-}(aq) + 2H^+(aq) = H_2O(l) + CO_2(g)$$

When sulphites are treated with acids they behave in a similar manner to carbonates. Sulphurous acid, H_2SO_3, is unstable and decomposes into water and sulphur dioxide gas, $e.g.$

$$Na_2SO_3 + 2HCl = 2NaCl + H_2SO_3$$
$$H_2SO_3 = H_2O + SO_2$$
therefore $Na_2SO_3 + 2HCl = 2NaCl + H_2O + SO_2$

The formation of volatile compounds (acids or covalent bases)

When concentrated sulphuric acid is added to solid sodium chloride, covalent hydrogen chloride is formed as a gas which escapes from the mixture:

$$H_2SO_4 + NaCl = NaHSO_4 + HCl,$$
$i.e.$ $H^+ + Cl^- = HCl(g)$

A similar reaction occurs when concentrated sulphuric acid is warmed with a nitrate; here nitric acid vapour escapes, $e.g.$

$$H_2SO_4 + NaNO_3 = NaHSO_4 + HNO_3,$$
$i.e.$ $H^+ + NO_3^- = HNO_3(g)$

The formation of weak acids

Weak acids, such as ethanoic acid, are often displaced from their salts by strong acids, such as hydrochloric acid or sulphuric acid. The reason is that for weak acids there is a particularly strong attraction between hydrogen ions and negatively charged ions; that is why weak acids have only a slight tendency to ionise in solution. Consequently when, say, sodium ethanoate is treated with hydrochloric acid there is a strong attraction between ethanoate ions and hydrogen ions, and covalent molecules of ethanoic acid are formed in solution:

$$\underbrace{CH_3COO^- + Na^+}_{\text{sodium ethanoate}} + \underbrace{H^+ + Cl^-}_{\text{hydrochloric acid}} = \underbrace{CH_3COOH}_{\text{ethanoic acid}} + \underbrace{Na^+ + Cl^-}_{\text{sodium chloride}}$$

i.e. $CH_3COO^-(aq) + H^+(aq) = CH_3COOH(aq)$

The reactions of carbonates or sulphites with mineral acids also fall into this category, but in these cases the acid which is formed decomposes and does not remain in solution.

Oxidation-reduction (redox)

Oxidation and reduction reactions have previously been encountered in Chapter 10.

If we look at oxidation reactions in more detail, we see that whenever a substance is oxidised it loses electrons. For example, we saw in Chapter 10 that when magnesium burns in air it gains oxygen and is oxidised to magnesium oxide:

$$2Mg + O_2 = 2MgO$$

The product, magnesium oxide, is electrovalent, and consists of magnesium ions, Mg^{2+}, and oxide ions, O^{2-}. Thus, the essential change that magnesium suffers in this reaction is *electron loss* as atoms are converted into ions:

$$Mg = Mg^{2+} + 2e^-$$

The following changes provide further examples of oxidation, in each of which the ion on the left is oxidised by the loss of one or more electrons:

$$Fe^{2+} = Fe^{3+} + e^-$$
$$Cl^- = Cl + e^-$$

Oxidation is therefore defined as electron loss. The 'o' in *o*xidation and 'o' in *lo*ss can assist in remembering this.

We saw in Chapter 10 that reduction is the reverse of oxidation. Consequently, because oxidation involves electron loss, reduction must entail electron gain. For instance, when copper(II) oxide is reduced to copper according to the equation:

$$CuO + H_2 = Cu + H_2O,$$

copper(II) ions in the copper(II) oxide gain electrons to give atoms of copper:

$$Cu^{2+} + 2e^- = Cu$$

The following conversions provide further examples of reduction. In each case the ion on the left-hand side is reduced by gaining one or more electrons.

$$Fe^{3+} + e^- = Fe^{2+}$$
$$Sn^{4+} + 2e^- = Sn^{2+}$$

Electron transfer

Substances do not become oxidised simply by emitting electrons to their surroundings. Similarly, substances never become reduced by collecting stray electrons from the surroundings. In the case of oxidation, the electrons which are lost are transferred to another substance which therefore becomes reduced. A chemical species, *i.e.* an atom, molecule or ion, cannot undergo oxidation unless there is another species present which, by gaining the lost electrons, can be reduced. Likewise, in the case of reduction, the electrons must always be gained from another chemical species which thereby becomes oxidised.

Oxidation therefore never occurs without reduction, and reduction never occurs without oxidation. Because of this, it is usually preferable to describe a chemical reaction of this type as a *redox, i.e.* reduction-oxidation, reaction, rather than just 'oxidation' or 'reduction'. The latter terms are best reserved for the changes affecting single species.

Every redox reaction therefore involves the transfer of electrons from the species which is oxidised to the species which is reduced.

To illustrate these principles, let us consider the so-called 'thermite reaction' between aluminium and iron(III) oxide to form aluminium oxide and iron:

$$Fe_2O_3 + 2Al = Al_2O_3 + 2Fe$$

The two oxides are electrovalent, *i.e.* $(Fe^{3+})_2 (O^{2-})_3$ and $(Al^{3+})_2 (O^{2-})_3$ respectively, and we can rewrite the equation to show this:

$$(Fe^{3+})_2 (O^{2-})_3 + 2Al = (Al^{3+})_2 (O^{2-})_3 + 2Fe$$

The oxide ions are spectator ions, not involved in the redox reaction. They can be cancelled out to give the following ionic equation:

$$Fe^{3+} + Al = Al^{3+} + Fe \qquad [1]$$

From this we can see that each aluminium atom is converted into an aluminium ion by the loss of three electrons:

$$Al = Al^{3+} + 3e^- \qquad [2]$$

Hence, aluminium is oxidised. The electrons are transferred to iron(III) ions which are thereby reduced to iron atoms:

$$Fe^{3+} + 3e^- = Fe \qquad [3]$$

To summarise, this redox reaction involves the transfer of three electrons from every aluminium atom to an iron(III) ion.

Equations [2] and [3] are known as *ionic half-equations*, because each of them represents half the redox reaction. If we add them together the electrons cancel out and we get the ionic equation [1] for the complete change.

Oxidising agents and reducing agents

In the thermite reaction we can regard the aluminium as bringing about the reduction of the iron(III) oxide, and the iron(III) oxide as bringing about the oxidation of the aluminium. A substance, such as aluminium in this example, which brings about the reduction of another substance by supplying it with electrons, is known as a *reducing agent* or a *reductant*. A substance, such as iron(III) oxide in the thermite reaction, which brings about the oxidation of another substance by accepting electrons from it, is termed an *oxidising agent* or an *oxidant*.

Because they give up electrons, reducing agents are always oxidised in a redox reaction (*e.g.* Al to Al_2O_3 in the above example). Oxidising agents, because they accept electrons, always become reduced (*e.g.* Fe_2O_3 to Fe).

As another example, let us examine the reaction between iron(II) chloride and chlorine to form iron(III) chloride:

$$2FeCl_2 + Cl_2 = 2FeCl_3$$

If we write each electrovalent compound in its ionic form, *i.e.*

$$2(Fe^{2+} + 2Cl^-) + Cl_2 = 2(Fe^{3+} + 3Cl^-),$$

with *reduction* over the right portion and *oxidation* under the left portion.

we see that iron(II) ions are converted to iron(III) ions:

$$Fe^{2+} = Fe^{3+} + e^-$$

This is oxidation; hence iron(II) chloride is the reducing agent. What is the oxidising agent? A further glance at the equation shows that there are four chloride ions on the left-hand side but six on the right. The two additional chloride ions are obtained by the reduction of chlorine atoms, which in turn are obtained from chlorine molecules:

$$Cl_2 + 2e^- = 2Cl^-$$

Chlorine is thus the oxidising agent, and this redox reaction involves the transfer of electrons from iron(II) ions to chlorine molecules.

Common oxidising agents and reducing agents

Oxidising agents Any chemical species (*i.e.* any atom, molecule or ion) which can gain electrons and become reduced is capable of acting as an oxidising agent. The more readily electrons are accepted, the more powerful is the oxidising action. Species such as H_2O, OH^- and $SO_4{}^{2-}$, in which all the elements are in a relatively stable state, do not readily accept electrons and are of no practical value as oxidising agents.

Non-metals, such as oxygen and chlorine, serve as oxidising agents because they have a strong tendency to receive electrons and form negatively charged ions:

$$O_2 + 4e^- = 2O^{2-}$$
$$Cl_2 + 2e^- = 2Cl^-$$

Certain acids and ions that contain an element in a high and relatively unstable valency state also behave as oxidising agents. Examples are concentrated nitric acid, concentrated sulphuric acid, dichromate ions and permanganate ions. Nitric acid and sulphuric acid are discussed later in this chapter, but dichromate and permanganate ions must be left until a higher level of study.

Reducing agents In principle, a reducing agent is any chemical species which is capable of losing electrons, thereby becoming oxidised. To be of real use it must be highly reactive, *i.e.* it must have a strong tendency to lose electrons. Metals near the top of the electrochemical series (see p. 193) satisfy this condition, for they readily form positive ions by the loss of electrons. Sodium and zinc are both well known laboratory reducing agents:

$$Na = Na^+ + e^-$$
$$Zn = Zn^{2+} + 2e^-$$

Ions and molecules which contain an element in a low and relatively unstable valency state can also be employed. Of particular note are the tin(II) and iron(II) ions, which become oxidised to tin(IV) and iron(III) ions respectively:

$$Sn^{2+} = Sn^{4+} + 2e^-$$
$$Fe^{2+} = Fe^{3+} + e^-$$

Substances which act as both oxidants and reductants If a substance can be both oxidised and reduced, then it will behave as either a reducing agent or an oxidising agent, depending upon what reagent is added. When treated with an oxidising agent more powerful than itself the substance will become oxidised and thus act as a reducing agent; conversely, in the presence of a more powerful reducing agent it will act as an oxidising agent.

An example is provided by hydrogen peroxide, H_2O_2, which is oxidised to oxygen by potassium permanganate (a relatively powerful oxidising agent) but reduced to water by sodium sulphite (a relatively powerful reducing agent). In the first case hydrogen peroxide behaves as a reducing agent, and in the second case as an oxidising agent.

Simple tests for oxidising agents and reducing agents

There is no absolute test for either type of substance, but it is possible to obtain a good insight into the nature of a compound by applying the following tests.

Oxidising agents

Many oxidising agents produce a deep blue-black colour when treated with an acidic solution of potassium iodide and starch. This is because the iodide ion is readily oxidised to iodine:

$$2I^- = I_2 + 2e^-$$

If starch is present, the iodine combines with the starch to give a blue-black coloured compound.

For convenience, starch-iodide papers are available. These are strips of absorbent paper which have been soaked in a solution of potassium iodide and starch and allowed to dry. When dipped into an acidified solution of an oxidising agent, starch-iodide paper turns a deep blue-black colour.

Some of the more powerful oxidising agents oxidise chloride ions, in acidic solution, to chlorine:

$$2Cl^- = Cl_2 + 2e^-$$

When such an oxidising agent is carefully warmed with concentrated hydrochloric acid in a test tube, chlorine gas is given off. The gas can be recognised by its pale green colour, and by its action on blue litmus paper or starch-iodide paper. A strip of moist blue litmus paper, held over the mouth of the tube, turns red and is then bleached if chlorine is present. Starch-iodide paper turns blue-black, because chlorine oxidises the iodide ions to iodine:

$$Cl_2 + 2I^- = I_2 + 2Cl^-$$

Most oxidising agents oxidise iron(II) ions to form iron(III) ions in acidic solution, and this forms the basis of another quick test. To a fresh solution of iron(II) sulphate in dilute sulphuric acid is added a small amount of the suspected oxidising agent. The presence of iron(III) ions can be detected by adding a few drops of a solution of either potassium hexacyanoferrate(II) or potassium thiocyanate. The former gives a deep blue precipitate with iron(III) ions, while the latter produces a red colouration. Thus, if the suspect substance yields a blue precipitate or a red colouration, depending on which reagent is used, it is most probably an oxidising agent. It is always advisable to check a portion of the original iron(II) sulphate solution with potassium hexacyanoferrate(II) or potassium thiocyanate to ensure that it contains no detectable iron(III) ions.

Reducing agents

Potassium dichromate, $K_2Cr_2O_7$, is a common oxidising agent which is reduced in acidic solution by many reducing agents. A distinctive colour change occurs when this happens:

$$K_2Cr_2O_7 \xrightarrow{\text{reduced to}} Cr^{3+}$$
orange   green

A solution of potassium dichromate, acidified with dilute sulphuric acid, is commonly used to test for gaseous reducing agents, *e.g.* sulphur dioxide or hydrogen sulphide. A few drops of the acidified dichromate solution are placed on a filter paper, which is then exposed to the suspect gas. The orange spot on the paper turns green on reduction.

Liquid or solid reducing agents can be detected by gently warming the suspect compound in a test tube with a dilute solution of potassium dichromate and sulphuric acid. The same colour change indicates that the material under test is a reducing agent.

Distinction between redox reactions and other types of reactions

As we have seen, a redox reaction involves electron transfer from a reducing agent to an oxidising agent. Transfer of a different kind occurs in neutralisation reactions. Consider, for example, the reaction between hydrochloric acid and sodium hydroxide:

$HCl + NaOH = NaCl + H_2O$

$i.e.$ $H^+(aq) \, Cl^- + Na^+OH^-(aq) = Na^+ \, Cl^- + H_2O$

Cancellation of spectator ions gives the ionic equation:

$H^+(aq) + OH^-(aq) = H_2O$

When we recall that a 'hydrogen ion' is a hydrated proton (see p. 140), we can see that essentially all that happens is that a proton from the acid is transferred to a hydroxide ion from the base to give a molecule of water. All acid-base reactions involve proton transfer from an acid to a base.

In double decomposition reactions there is movement of neither electrons nor protons but of ions. This type of reaction can in fact be defined as an interchange of ions. To summarise,

oxidation-reduction involves electron transfer,
neutralisation involves proton transfer,
double decomposition involves ion transfer.

The oxidising properties of acids

All acids have the ability to convert metals into salts, such as chlorides, sulphates and nitrates. These salts are electrovalent and contain metal ions, M^{n+}, where n is the valency of the metal. Consequently, when an acid attacks a metal, atoms of that metal become oxidised by electron loss:

$M = M^{n+} + ne^-$

Often, hydrogen ions from the acid are reduced by electron gain so that hydrogen gas is formed:

$2H^+(aq) + 2e^- = H_2$

Thus, all acids are oxidising agents. However, some are particularly good oxidising agents, because their negative ions or their molecules are reduced very easily. Nitric acid is a good example. In solution it provides the nitrate ion, NO_3^-, which is easily reduced, $e.g.$ to nitrogen dioxide, NO_2, which is given off as a brown gas:

$NO_3^- + 2H^+(aq) + e^- = NO_2 + H_2O$

Acids of this sort are known as *oxidising acids*. Apart from nitric acid, the only oxidising acid of importance is concentrated sulphuric acid, whose molecules are easily reduced, *e.g.* to sulphur dioxide:

$$2H_2SO_4 + 2e^- = SO_2 + SO_4^{2-} + 2H_2O$$

Acids of the first sort, whose negative ions or molecules are not readily reduced, can be termed *weakly oxidising acids*. Examples are hydrochloric acid, dilute sulphuric acid, and ethanoic acid.

The action of acids on the elements

Metals Sodium and potassium react explosively with acids, and such reactions should not be attempted. Calcium reacts vigorously but not explosively. In this section, however, we are concerned only with the reactions of common metals, *i.e.* magnesium, zinc, aluminium, iron and copper, with nitric acid, hydrochloric acid and sulphuric acid. In all cases the metal is oxidised to a salt, but the reduction product of the acid depends on whether an oxidising or weakly oxidising acid is used.

With weakly oxidising acids
Metals which are above hydrogen in the electrochemical series (see p. 193), *e.g.* magnesium, zinc, aluminium and iron, react with hydrochloric acid and dilute sulphuric acid to produce hydrogen and metal chlorides and sulphates respectively, *e.g.*

$$Zn + 2HCl = ZnCl_2 + H_2$$

We can say that zinc 'displaces' hydrogen from hydrochloric acid.

The reaction can be understood in terms of electron transfer from zinc to hydrogen ions as explained above, *i.e.*

$$Zn = Zn^{2+} + 2e^-$$
$$2H^+(aq) + 2e^- = 2H$$
$$2H = H_2$$

With oxidising acids
With the exception of the reaction between magnesium and 2 per cent nitric acid, hydrogen is never formed when metals react with oxidising acids. This is because oxidising acids, or their negative ions, are reduced in preference to hydrogen ions. Aluminium, iron and copper, for example, reduce hot concentrated sulphuric acid to sulphur dioxide, *e.g.*

$$Cu + 2H_2SO_4 = CuSO_4 + SO_2 + 2H_2O$$

Zinc and magnesium react with concentrated sulphuric acid to form hydrogen sulphide, *e.g.*

$$4Zn + 5H_2SO_4 = 4ZnSO_4 + H_2S + 4H_2O$$

The reactions between metals and nitric acid are complicated by the fact that the product to which the nitric acid is reduced depends on the concentration of the acid, the temperature and the nature of the metal. Some of the reduction products of nitric acid are nitrogen dioxide, NO_2, nitrogen oxide, NO, and the ammonium ion, NH_4^+. Reactive metals, such as magnesium or zinc, reduce nitric acid to the ammonium ion, *e.g.*

$$4Zn + 10HNO_3 = 4Zn(NO_3)_2 + NH_4NO_3 + 3H_2O$$

Copper and *dilute* nitric acid produce nitrogen oxide,

$$3Cu + 8HNO_3 = 3Cu(NO_3)_2 + 2NO + 4H_2O,$$

but with the concentrated acid nitrogen dioxide is formed:

$$Cu + 4HNO_3 = Cu(NO_3)_2 + 2NO_2 + 2H_2O$$

Iron, like copper, reacts with dilute nitric acid to form nitrogen oxide, but with the concentrated acid no reaction occurs. Nitric acid of any concentration does not attack aluminium.

Non-metals There is never any chemical reaction between non-metals and weakly oxidising acids. Carbon, sulphur, nitrogen, phosphorus, chlorine, etc. are not attacked by hydrochloric acid or dilute sulphuric acid. Oxidising acids, however, oxidise many non-metals to their oxides or oxoacids, *e.g.*

$$S + 2H_2SO_4 = 3SO_2 + 2H_2O$$

Industrial redox reactions

Redox reactions are commonly used in industry to isolate metals from their ores. *Ores* are mineral substances, *i.e.* compounds which occur naturally, from which metals can be profitably extracted. They are often contaminated by other minerals or soil, and some purification may be necessary before metals can be obtained from them.

Many ores are metal oxides, *e.g.* Fe_2O_3, sulphides, *e.g.* ZnS and PbS, or carbonates, *e.g.* $ZnCO_3$ and $FeCO_3$. It is not easy to extract metals directly from sulphides and carbonates, and ores of this kind are usually converted to oxides by roasting them in air, *e.g.*

$$2PbS + 3O_2 = 2PbO + 2SO_2$$
$$ZnCO_3 = ZnO + CO_2$$

Metal oxides are ionic, containing metal ions and oxide ions, and in order to convert them to metals they must be reduced. For example, the conversion of zinc ions, in ZnO, to zinc is a reduction process:

$$Zn^{2+} + 2e^- = Zn$$

A reducing agent is therefore necessary to convert a metal oxide to a metal.

For industrial purposes it is important that the reducing agent is cheap and readily available. Two such substances are carbon and carbon monoxide. At room temperature neither is effective as a reducing agent, but at temperatures greater than about 700 °C (973 K) they will reduce many metal oxides to metals. Iron, for example, is obtained from iron(III) oxide by heating it with carbon in a vertical furnace called a *blast furnace* (see Fig. 12.3). The furnace is loaded with iron(III) oxide, coke (*i.e.* carbon) and limestone (*i.e.* calcium carbonate) at the top, and air is forced in at the base.

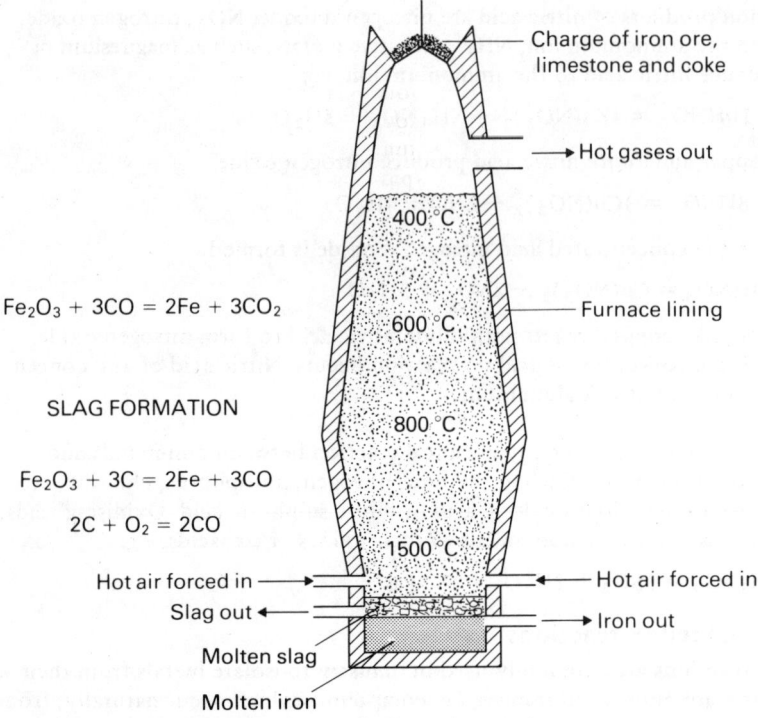

Fig. 12.3 The blast furnace

Many reactions occur in the blast furnace, but the main ones are as follows. Some of the iron(III) oxide is reduced by carbon:

$$Fe_2O_3 + 3C = 2Fe + 3CO$$

but most of it is reduced by carbon monoxide:

$$Fe_2O_3 + 3CO = 2Fe + 3CO_2$$

The carbon monoxide required for this is formed by the partial combustion of coke at the base of the furnace:

$$2C + O_2 = 2CO$$

The purpose of the limestone is to form a slag with the earthy impurities in the iron ore. Molten iron and slag are drawn off from time to time at the bottom of the furnace.

The oxides of lead and zinc are reduced with carbon in a similar manner. Some important metals, notably aluminium, cannot be produced by reduction of their ores with carbon or carbon monoxide. Aluminium is manufactured by electrolysis (see p. 188), which is another means of bringing about reduction.

The hazards of redox reactions

Some redox reactions are strongly exothermic, *i.e.* much heat is given out when they take place. For example, in the thermite reaction (see above), so much heat is released that iron is produced in the molten state. While this can be useful, *e.g.* for welding steel in inaccessible places, it can also present dangers. If, for instance, the reaction is carried out in a crucible standing on a wooden bench, the heat liberated can cause the crucible to burn a hole right through the bench!

Dangers arise particularly in two situations.

With powerful oxidising agents or powerful reducing agents Some of the strongest oxidising agents in common use are concentrated nitric acid, concentrated sulphuric acid, potassium permanganate and potassium dichromate. These substances must always be kept off the skin, clothing, paper and wood, for they are capable of causing serious damage. If they are accidentally spilt on the skin the first aid treatment is water; plenty of it and as quickly as possible!

The most powerful reducing agent in common use is sodium. This is so strong that it is able to reduce water to hydrogen, and the hydrogen can then ignite. Extreme care is needed in handling this material.

Industrial safety goggles must always be worn when handling dangerous chemicals.

With redox reactions which lead to the sudden release of gas Some redox reactions are explosive, in that they lead to the sudden release of relatively large quantities of hot gas. A well known example concerns gunpowder, which is a mixture of potassium nitrate, charcoal and sulphur. Potassium nitrate is an oxidising agent, while charcoal and sulphur are both reducing agents. When gunpowder explodes, potassium nitrate rapidly oxidises charcoal (which is mainly carbon) to carbon dioxide gas, and sulphur to sulphur dioxide gas. If the gunpowder is contained in a cartridge, the sudden release of these gases is sufficient to propel a bullet from the cartridge at high velocity.

Explosive reactions should never be performed in student laboratories.

Decomposition

A *decomposition* reaction is one in which a compound is decomposed, *i.e.* split up, into simpler substances. If the reaction is brought about by the action of heat, and if it is irreversible, *i.e.* permanent, it is described as a *thermal decomposition*. Examples include the decomposition of potassium chlorate and ammonium nitrate by the action of heat:

$$2KClO_3 = 2KCl + 3O_2$$
$$NH_4NO_3 = N_2O + 2H_2O$$

These reactions are irreversible. The products do not, under any circumstances, recombine to form the original substances.

A thermal decomposition which is reversible is called a *thermal dissociation*. For example, if calcium carbonate is heated in a sealed container it is partly split up into calcium oxide and carbon dioxide. The reaction, however, is reversible, because the two products combine together to reform calcium carbonate:

$$CaCO_3 = CaO + CO_2$$

Other examples include the dissociation of ammonium chloride into ammonia and hydrogen chloride, and the dissociation of dinitrogen tetraoxide to nitrogen dioxide:

$$NH_4Cl = NH_3 + HCl$$
$$N_2O_4 = 2NO_2$$

Summary

At the conclusion of this chapter, you should be able to:

1. select examples of neutralisation, redox or double decomposition reactions from a given set of equations,
2. perform acid-base titrations,
3. use volumetric glassware accurately,
4. define 'titre' as the volume of solution delivered from a burette in reaching the end-point of a titration,
5. describe the action of acids and bases on the colours of common indicators,
6. calculate the concentration (molarity) of a solution from the results of a titration,
7. state that double decomposition reactions involve an exchange of ions between two reactants,
8. recognise that double decomposition reactions proceed to completion only if a pair of oppositely charged ions is permanently removed from solution,
9. define 'redox' in terms of electron transfer,
10. identify in a given redox reaction which reactant undergoes oxidation and which reduction,
11. perform simple tests to identify the presence of oxidising or reducing agents,
12. state the difference between redox, neutralisation and double decomposition reactions,
13. describe the action of acids on metals in terms of oxidation of the metal and reduction of the acid,
14. recognise the hazards associated with reactions between acids and certain metals,
15. identify important industrial redox reactions, especially in the extraction of metals,
16. recognise that redox systems are potentially hazardous,
17. define 'thermal decomposition' and 'thermal dissociation'.

Questions

Classify each of the reactions shown in questions 1–6, by selecting the most appropriate term from the list below.

A	Neutralisation	C	Oxidation-reduction
B	Double decomposition	D	Decomposition

1. $MnO_2 + 4HCl = Cl_2 + MnCl_2 + 2H_2O$
2. $Ca + 2H_2O = Ca(OH)_2 + H_2$
3. $2HCl + Ca(OH)_2 = CaCl_2 + 2H_2O$
4. $H_2SO_4 + KNO_3 = HNO_3 + KHSO_4$
5. $ZnCO_3 = ZnO + CO_2$
6. $CaCO_3 + 2HCl = CaCl_2 + H_2O + CO_2$

   In questions 7 and 8 the following relative atomic masses are required:
   $H = 1, C = 12, N = 14, O = 16, Na = 23, S = 32, Cl = 35.5$.

7. Calculate the volume of sodium hydroxide solution of concentration 0.100 mol dm^{-3} (0.100 M) required to neutralise:
   (a) 10.0 cm^3 of hydrochloric acid of concentration 0.100 mol dm^{-3} (0.100 M),
   (b) 25.0 cm^3 of hydrochloric acid of concentration 0.105 mol dm^{-3} (0.105 M),
   (c) 23.8 cm^3 of sulphuric acid of concentration 0.050 mol dm^{-3} (0.050 M),
   (d) a solution containing 0.02 mol of nitric acid,
   (e) a solution containing 1 g of pure nitric acid,
   (f) 15.0 cm^3 of a solution containing 10.0 g dm^{-3} of pure nitric acid.

8. The following experiments were performed to determine the exact concentration of a solution of sodium hydroxide, known to have an *approximate* concentration of 0.1 mol dm^{-3} (0.1 M).
   First, a standard solution of sodium carbonate was prepared by dissolving 1.300 g of anhydrous sodium carbonate in distilled water so as to make 250 cm^3 of solution.
   25.0 cm^3 of this solution was pipetted into a conical flask and titrated against hydrochloric acid, with methyl orange as indicator. 26.80 cm^3 of hydrochloric acid was required to reach the end-point. Finally, with the same hydrochloric acid still in the burette, 22.75 cm^3 of hydrochloric acid was required in a titration with 25.0 cm^3 of the sodium hydroxide solution. Calculate:
   (a) the concentration (molarity) of the sodium carbonate,
   (b) the concentration (molarity) of the hydrochloric acid,
   (c) the concentration (molarity) of the sodium hydroxide,
   (d) the mass concentration (*i.e.* concentration in g dm^{-3}) of the sodium hydroxide.

   With reference to question 8, the glass apparatus could be rinsed before use with

A	distilled water	C	hydrochloric acid
B	sodium carbonate solution	D	sodium hydroxide solution

   Select from this list the most suitable material for rinsing each item of apparatus shown in questions 9–13.

9. The graduated flask.
10. The burette for the first titration.
11. The pipette for the first titration.
12. The conical flask for the first titration.
13. The conical flask for the second titration (*i.e.* NaOH/HCl).
14. Which of the following statements about oxidation and reduction are true and which are false?
    (*a*) Oxidation and reduction never occur together.
    (*b*) Oxidation or reduction always involves a change in the valency of an element.
    (*c*) Oxidation is defined as electron loss.
    (*d*) An oxidising agent becomes oxidised during a redox reaction.
    (*e*) A reducing agent gains electrons during a redox reaction.
    (*f*) When a substance becomes reduced it gains electrons from the atmosphere.
    (*g*) Redox reactions are never strongly exothermic.
    (*h*) Oxidation must involve oxygen or an oxygen-containing substance.
    (*i*) Redox reactions always involve the transfer of oxygen atoms from the oxidising agent to the reducing agent.
    (*j*) Redox reactions always involve the transfer of electrons from the oxidising agent to the reducing agent.
    (*k*) Certain substances can behave as oxidising agents in some reactions and reducing agents in others.
15. In which of the following equations for redox reactions are the oxidising agents printed on the extreme left?
    A $2Na + 2H_2O = 2NaOH + H_2$
    B $Zn + H_2SO_4 = ZnSO_4 + H_2$
    C $Zn + 2H_2SO_4 = ZnSO_4 + SO_2 + 2H_2O$
    D $2Na + Cl_2 = 2NaCl$
    E $C + O_2 = CO_2$
    F $ZnO + C = Zn + CO$
    G $CuO + H_2 = Cu + H_2O$
    H $2FeCl_3 + SnCl_2 = 2FeCl_2 + SnCl_4$
    For each of the questions 16–20, select from the list below the number of electrons required to balance the ionic half-equation, *i.e.* select the correct value of *n*.
    A   1   C   3
    B   2   D   4
16. $Mg = Mg^{2+} + ne^-$
17. $Sn^{2+} = Sn^{4+} + ne^-$
18. $Br_2 = 2Br^- - ne^-$
19. $NO_3^- + 2H^+ = NO_2 + H_2O - ne^-$
20. $H_2O_2 = O_2 + 2H^+ + ne^-$

In questions 21–24, select the most appropriate answer, labelled A, B, C or D.

21. Neutralisation (*i.e.* acid-base) reactions involve the transfer of:
    A protons,
    B neutrons,
    C electrons,
    D ions.

22. The colour change of methyl orange on progressing from acidic to basic solution is:
    A orange to pale pink,
    B colourless to orange,
    C red to yellow,
    D yellow to red.

23. Zinc displaces hydrogen from:
    A dilute nitric acid,
    B dilute sulphuric acid,
    C concentrated nitric acid,
    D concentrated sulphuric acid.

24. An unknown substance turns starch-iodide paper blue. This shows that the substance is:
    A an oxidising agent,
    B a reducing agent,
    C an acid,
    D a base.

Chapter 13

Salts

A *salt* is a compound formed when some or all of the acidic hydrogen atoms in the molecule of an acid are replaced by metal atoms or the ammonium group of atoms, NH_4. Since all acids ionise in solution to produce hydrogen ions, and since the great majority of salts are ionic, we can define a salt as **a compound formed by replacing all or some of the hydrogen ions of an acid by metal ions or ammonium ions.** Thus, the replacement of hydrogen ions from hydrochloric acid by sodium ions, calcium ions or iron(III) ions gives the salts $NaCl$, $CaCl_2$ or $FeCl_3$ respectively.

If all of the hydrogen ions are not replaced then *acid salts* may be produced. For example, replacement of both the hydrogen ions from a sulphuric acid molecule by sodium ions gives the normal salt, sodium sulphate, Na_2SO_4. However, if only one of the hydrogen ions is replaced, we obtain a salt which still contains replaceable hydrogen. This is the acid salt, sodium hydrogen-sulphate, $NaHSO_4$.

Solubility of salts

Salts can be divided into two broad categories, 'soluble' and 'insoluble', according to the extent to which they dissolve in water (see Table 13.1).

A knowledge of the solubility of salts is essential when considering their preparation, especially by double decomposition reactions, and Table 13.1 will be found to be useful in this respect. There is, however, no sharp dividing line between soluble and insoluble compounds, and some of the salts (*e.g.* $CaSO_4$) listed in Table 13.1 as insoluble in water are in fact slightly soluble. Furthermore, it should be noted that salts which are commonly said to be insoluble in water do dissolve to a very small extent.

Table 13.1 Soluble and insoluble salts

Soluble salts	Insoluble salts
All nitrates	
Most salts of sodium, potassium and ammonia	
Chlorides, except	$AgCl$, $PbCl_2$, Hg_2Cl_2 and $CuCl$
Sulphates, except	$PbSO_4$, $CaSO_4$, $SrSO_4$ and $BaSO_4$
Carbonates of sodium, potassium and ammonia	All other carbonates

Preparation of salts

Several methods are available, but only the principal ones will be discussed here.

Neutralisation

A solution of a salt is obtained whenever an acid is neutralised by a base (see Chapter 11).

Acid plus an insoluble base

Most metal oxides and hydroxides are insoluble in water, but because they are basic they will dissolve in acids to give salts. Small portions of the base are added with stirring to the acid, which should preferably be heated to speed up the reaction, until no more will dissolve. A solution of copper(II) sulphate can be prepared in this way by adding either copper(II) oxide or copper(II) hydroxide to a solution of hot dilute sulphuric acid:

$$CuO + H_2SO_4 = CuSO_4 + H_2O$$
$$Cu(OH)_2 + H_2SO_4 = CuSO_4 + 2H_2O$$

When no more base dissolves, the acid is neutralised and the reaction is complete. The excess base is then removed from the salt solution by filtration (see p. 9) and the filtrate (*i.e.* the solution) is crystallised (see p. 88) to obtain crystals of the salt.

Acid plus a soluble electrovalent base

The problem here is to know when sufficient base has been added to the acid to just neutralise it. An excess of base must be avoided, for otherwise crystallisation yields not only the salt but also the original base as an impurity.

We overcome this difficulty by carrying out a titration (see p. 151). For example, to prepare sodium sulphate from a solution of sodium hydroxide and dilute sulphuric acid, we can pipette, say, 25 cm^3 of sulphuric acid into a conical flask and add methyl orange solution dropwise until a red colour is obtained. The acid is titrated against sodium hydroxide solution from a burette until the indicator changes colour to orange. The volume of sodium hydroxide required is noted. At this point sufficient sodium hydroxide has been added to just neutralise the acid to form sodium sulphate:

$$H_2SO_4 + 2NaOH = Na_2SO_4 + 2H_2O$$

The solution obtained in this way contains sodium sulphate contaminated by methyl orange. However, if the experiment is repeated, using exactly the same quantities of sulphuric acid and sodium hydroxide as before, but leaving out the methyl orange, the resulting solution can be crystallised to yield pure sodium sulphate.

If the experiment is repeated with *half* the volume of sodium hydroxide, but with the same volume of sulphuric acid, the acid salt, sodium hydrogen-sulphate, is obtained:

$$H_2SO_4 + NaOH = NaHSO_4 + H_2O$$

Similar experiments can be performed with any acid and any soluble base.

Acid plus ammonia

In the preparation of an ammonium salt from aqueous ammonia and an acid the titration can be omitted. All that is necessary is to ensure that an excess of ammonia is added to the acid. This is achieved by dipping a piece of red litmus paper into the solution after the aqueous ammonia has been added. When an excess of ammonia is present the litmus paper turns blue. Crystallisation of the solution produces crystals of the pure ammonium salt, because the excess ammonia is lost together with water as the solution is evaporated.

Double decomposition

Double decomposition, the principles of which were discussed in Chapter 12, is a very useful means of converting one salt into another.

Acid plus an insoluble carbonate

Carbonates are salts which resemble bases in that they have the ability to neutralise acids to form salts of those acids. The neutralisation procedure depends on whether the carbonate is soluble or insoluble, just the same as when a metal oxide or hydroxide is used.

An insoluble carbonate is added to the acid in small portions, with stirring, until no more carbon dioxide is given off, *i.e.* until fizzing stops, and an excess of insoluble carbonate is present. The excess carbonate is then filtered off and the solution is crystallised to produce the salt. Magnesium sulphate, for example, can be prepared by adding magnesium carbonate to dilute sulphuric acid until no more carbon dioxide is evolved and undissolved magnesium carbonate remains:

$$MgCO_3 + H_2SO_4 = MgSO_4 + CO_2 + H_2O$$

After filtration and crystallisation of the solution crystals of magnesium sulphate are obtained.

Acid plus a soluble carbonate

This method is essentially the same as that described above for acid plus a soluble base. The problem of knowing when the acid has been neutralised is again resolved by means of a titration. For example, to prepare potassium nitrate from potassium carbonate and nitric acid:

$$K_2CO_3 + 2HNO_3 = 2KNO_3 + CO_2 + H_2O,$$

we must first titrate the solutions against each other with methyl orange as the indicator. Once we know how much carbonate is required to produce a neutral solution, we can add this amount to the acid without using an indicator.

Precipitation reactions

Salts which are insoluble in water can be prepared by precipitation reactions. Lead(II) chloride, for example, is obtained as a white precipitate by mixing together a solution of a soluble lead(II) salt, *e.g.* lead(II) nitrate, and a solution of a chloride, *e.g.* sodium chloride:

$$Pb(NO_3)_2 + 2NaCl = PbCl_2 + 2NaNO_3$$
i.e. $Pb^{2+}(aq) + 2Cl^-(aq) = PbCl_2(s)$

When precipitation is complete, the lead(II) chloride is filtered off, washed and dried. Sodium nitrate remains in solution, and can be recovered by crystallising the filtrate.

All three types of double decomposition reaction are useful in the preparation of salts by precipitation. Besides the interaction of two salts, just described, we can use acid plus a salt, *e.g.*

$$BaCl_2 + H_2SO_4 = BaSO_4 + 2HCl$$
i.e. $Ba^{2+}(aq) + SO_4{}^{2-}(aq) = BaSO_4(s),$

or base plus a salt, *e.g.*

$$Ca(OH)_2 + Na_2CO_3 = CaCO_3 + 2NaOH$$
i.e. $Ca^{2+}(aq) + CO_3{}^{2-}(aq) = CaCO_3(s)$

In all cases, after filtration, the insoluble salt is washed and dried.

The action of an acid on a metal

All metals except the least reactive are chemically attacked by acids to form salts. The reactions, as we saw in Chapter 12, involve oxidation of the metal and reduction of the acid.

Salts are commonly prepared by the action of weakly oxidising acids on metals which lie above hydrogen in the electrochemical series. (These are the more reactive metals, see p. 194). For example, iron(II) sulphate is obtained by dissolving iron in warm dilute sulphuric acid:

$$Fe + H_2SO_4 = FeSO_4 + H_2$$

An excess of the metal is employed to ensure that all of the acid reacts. After the evolution of hydrogen has ceased, the solution is filtered and the filtrate is evaporated to produce crystals of the salt.

Thermal decomposition of salts

In general, chlorides and sulphates are resistant to thermal decomposition, but nitrates, ammonium salts and most carbonates break down on heating into simpler substances.

Nitrates

Sodium nitrate and potassium nitrate decompose to give the corresponding nitrite, together with oxygen, *e.g.*

$$2NaNO_3 = 2NaNO_2 + O_2$$

Most other metal nitrates break down to give the metal oxide and a mixture of two gases, namely nitrogen dioxide and oxygen, *e.g.*

$$2Cu(NO_3)_2 = 2CuO + 4NO_2 + O_2$$

Carbonates

The carbonates of sodium and potassium are thermally stable, but other metal carbonates decompose at red heat to give the metal oxide and carbon dioxide, *e.g.*

$$CaCO_3 = CaO + CO_2$$

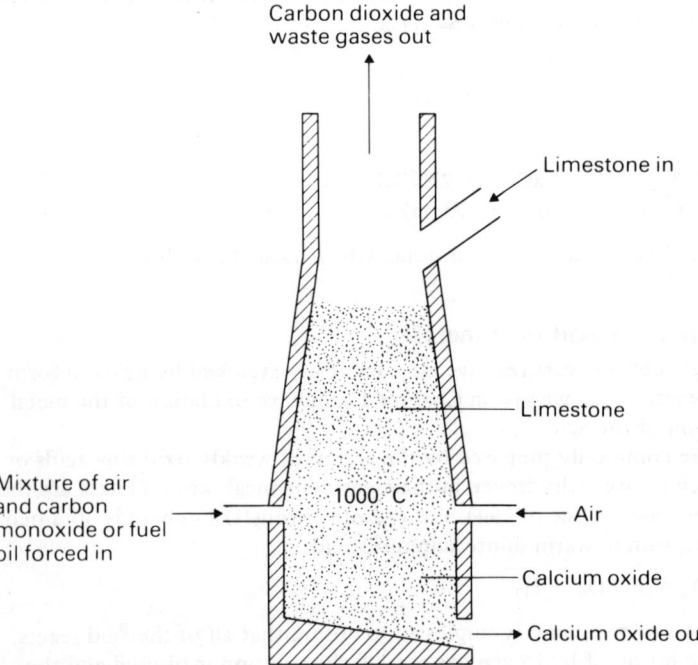

Fig. 13.1 A lime kiln

The decomposition of calcium carbonate in the form of limestone is particularly important in industry. Heating is carried out in 'limekilns' (see Fig. 13.1) to produce calcium oxide, 'quicklime'. The quicklime is treated with water to give calcium hydroxide, 'slaked lime', which is used mainly by farmers to neutralise excess acid in the soil.

$$CaO + H_2O = Ca(OH)_2$$

Ammonium salts

Ammonium salts on heating behave quite differently from metal salts. Many undergo thermal dissociation to give ammonia and the parent acid, *e.g.*

$$NH_4Cl = NH_3 + HCl$$

However, if the negative ion is an oxidising agent, it oxidises the ammonium ion to nitrogen or an oxide of nitrogen in an irreversible change, *e.g.*

$$NH_4NO_3 = N_2O + 2H_2O$$

In this example, dinitrogen oxide, N_2O, arises partly by the oxidation of the ammonium ion and partly by the reduction of the nitrate ion.

Tests on salts

As we have seen earlier, most salts are electrovalent compounds. Their crystals are composed of ions, as the following examples show.

Salt	Ions present
NaCl	Na^+ and Cl^-
K_2SO_4	K^+ and $SO_4{}^{2-}$
KNO_3	K^+ and $NO_3{}^-$
Na_2CO_3	Na^+ and $CO_3{}^{2-}$
NH_4Cl	$NH_4{}^+$ and Cl^-

When salts are dissolved in water they dissociate, *i.e.* the ions in the solid become separated from one another. The ions present in a solution of a salt are thus identical to those which exist in the solid. Sometimes it is necessary to test for the presence of ions in solution, and methods for detecting the more common ions are described below.

Ammonium ions

When a compound containing ammonium ions is treated with sodium hydroxide or potassium hydroxide and warmed, ammonia is evolved. The reaction is essentially one in which the ammonium ion loses a hydrogen ion to a hydroxide ion:

$$NH_4{}^+ + OH^- = NH_3 + H_2O$$

The ammonia which is evolved can be identified by its very distinctive smell and by the fact that it turns moist red litmus paper blue.

Chloride ions

If the solution is not already acidic, it is acidified by adding dilute nitric acid until a strip of blue litmus paper dipped into the solution turns red. A solution of silver nitrate is then added, and if chloride ions are present in the solution a white precipitate of silver chloride is obtained by a double decomposition reaction:

$$Ag^+(aq) + Cl^-(aq) = AgCl(s)$$

Further tests should be performed to prove conclusively that the precipitate is silver chloride. If the precipitate dissolves readily in aqueous ammonia, but not in concentrated nitric acid, it can be assumed to be silver chloride. This confirms the presence of chloride ions in the original solution.

Sulphate ions

If necessary, the solution should be acidified with sufficient dilute hydrochloric acid or dilute nitric acid. The formation of a white precipitate of barium sulphate when a solution of barium chloride is added is indicative of the presence of sulphate ions in the original solution:

$$Ba^{2+}(aq) + SO_4{}^{2-}(aq) = BaSO_4(s)$$

Nitrate ions

Brown ring test

A few cubic centimetres of the suspect nitrate solution are mixed with a similar volume of an aqueous iron(II) sulphate solution in a test tube. After cooling the mixture under the tap, the tube is held in a slanting position and about 2 cm^3 of concentrated sulphuric acid are carefully poured down the inside of the tube (see Fig. 13.2). **The tube must not be shaken as the acid is added.** The acid sinks to the bottom of the tube, and at the junction of the acid and the mixed solution a brown ring develops if the solution contains nitrate ions.

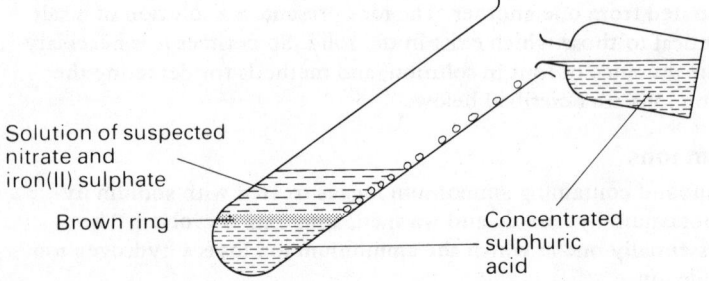

Solution of suspected nitrate and iron(II) sulphate

Brown ring

Concentrated sulphuric acid

Fig. 13.2 Correct technique for performing the brown ring test

The reactions that occur to produce the brown ring are rather complicated. First, nitrate ions, in the presence of an acid, are reduced by iron(II) ions to nitrogen oxide:

$$3Fe^{2+} + 4H^+(aq) + NO_3^- = 3Fe^{3+} + 2H_2O + NO$$

Nitrogen oxide then combines with unused iron(II) ions to form a brown coloured compound. This compound decomposes when heated; hence the need to cool the solution.

Devarda's alloy test

Devarda's alloy, a mixture of aluminium, copper and zinc, reduces nitrate ions in hot basic solution to ammonia, which can be detected by its smell or its action on litmus paper as described above.

The test is meaningless if ammonium ions are present in the solution under examination, because in basic solution they, too, liberate ammonia. Thus, if ammonium ions are present they must be removed by boiling the solution of the suspect nitrate in a test tube with sodium hydroxide solution until no more ammonia is evolved:

$$NH_4^+ + OH^- = NH_3 + H_2O$$

Afterwards, a small amount of Devarda's alloy is added and the basic solution is again boiled. The evolution of ammonia *at this stage* indicates the presence of nitrate ions in the original solution.

If ammonium ions are absent in the suspect nitrate, the initial boiling with sodium hydroxide before adding Devarda's alloy is omitted.

Carbonate ions

In the presence of mineral acid the carbonate ion forms carbon dioxide gas and water:

$$CO_3^{2-} + 2H^+(aq) = CO_2 + H_2O$$

Thus all carbonates, when treated with hydrochloric acid, nitric acid or sulphuric acid, evolve carbon dioxide. This provides a most convenient test for carbonates. If a compound *effervesces, i.e.* if it fizzes and gives off carbon dioxide, as soon as it is treated with an acid, it is a carbonate and contains carbonate ions. Hydrogencarbonate ions, HCO_3^-, also evolve carbon dioxide when treated with acids, but these are discussed at a higher level of study.

Any gas which is evolved in such a test must be proved to be carbon dioxide before any firm conclusion can be drawn. Carbon dioxide is readily identified by the fact that it affects lime-water, first turning it milky and later sending it clear again.

The simplest way to perform a carbonate test is to add dilute hydrochloric acid to the suspect material in a test tube. If effervescence occurs, a sample of the gas is collected in a teat pipette (see Fig. 13.3(a)). The bulb of the pipette must be squeezed *before* it is inserted in the tube.

The pipette is then withdrawn from the tube, and the gas is bubbled through a small amount of lime-water contained in another test tube (see Fig.

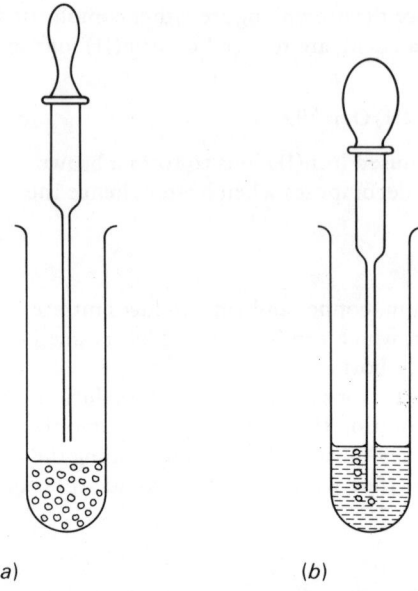

(a)   (b)

Fig. 13.3 Correct technique for testing for carbon dioxide

13.3(b)). A white precipitate or 'milkiness' is positive for the presence of carbon dioxide and hence carbonate ions in the original substance.

Lime-water is a saturated solution of the sparingly soluble base, calcium hydroxide, $Ca(OH)_2$. It reacts with carbon dioxide, an acidic gas, to form the salt, calcium carbonate:

$$Ca(OH)_2 + CO_2 = CaCO_3(s) + H_2O$$

It is the formation of insoluble calcium carbonate that causes the lime-water to turn milky.

In the presence of more carbon dioxide, calcium carbonate forms the soluble acid salt, calcium hydrogencarbonate; hence a clear solution is obtained from the milky lime-water:

$$CaCO_3 + H_2O + CO_2 = Ca(HCO_3)_2$$

Normally, the fact that a gas turns lime-water milky is sufficient to identify it as carbon dioxide.

Summary

At the conclusion of this chapter, you should be able to:
1. describe a salt as a substance formed when an acid is neutralised by a base, or when hydrogen in an acid is displaced by a metal,
2. state whether a given salt (chloride, sulphate, nitrate or carbonate) is soluble or insoluble in water,

3. list the principal methods of salt preparation as (a) neutralisation, (b) acid plus a carbonate, (c) precipitation, and (d) acid on a metal,
4. prepare salts in the laboratory by methods which involve titration, filtration or crystallisation,
5. describe the action of heat on common salts, *i.e.* chlorides, sulphates, nitrates, carbonates and ammonium salts,
6. perform tests to identify the presence of chloride, sulphate, nitrate, carbonate and ammonium ions.

Questions

Examine each of the suggested double decomposition reactions shown in questions 1–6, and select from the list below the phrase which best describes what will happen. Refer to the solubilities of salts in Table 13.1.

    A   proceeds from left to right because one of the products is insoluble
    B   proceeds from left to right because one of the products is volatile
    C   proceeds from right to left
    D   does not proceed to completion in either direction but gives a mixture of all four compounds

1. $CaCO_3 + 2NaCl = CaCl_2 + Na_2CO_3$
2. $Cu(NO_3)_2 + H_2SO_4 = CuSO_4 + 2HNO_3$
3. $MgCl_2 + 2NaNO_3 = Mg(NO_3)_2 + 2NaCl$
4. $Pb(NO_3)_2 + K_2SO_4 = PbSO_4 + 2KNO_3$
5. $FeCl_2 + 2AgNO_3 = 2AgCl + Fe(NO_3)_2$
6. $Al_2(SO_4)_3 + 6HCl = 2AlCl_3 + 3H_2SO_4$
7. Which of the following statements are true and which are false?
    (a) In testing for chloride ions, the solution must be acidified with dilute nitric acid.
    (b) In testing for sulphate ions, the solution must be acidified with dilute sulphuric acid.
    (c) An unknown compound, on treatment with dilute hydrochloric acid, gave off a gas which turned lime-water milky. This showed that the unknown compound was a carbonate or a hydrogencarbonate.
    (d) The brown ring test must be performed at the boiling temperature of the solution.

In questions 8–10 select the most appropriate answer, labelled A, B, C or D.

8. An unknown compound, on boiling with aqueous sodium hydroxide, gave off ammonia. When no more ammonia could be detected, a little Devarda's alloy was added and the mixture was boiled again. More ammonia was evolved. The unknown compound could have been:
    A   ammonium nitrate,
    B   ammonium chloride,
    C   any ammonium salt,
    D   any nitrate except ammonium nitrate.
9. An unknown solid gave off carbon dioxide when treated with dilute hydrochloric acid, but did not give off carbon dioxide on heating.

The compound was:

A   possibly a carbonate,

B   not a carbonate,

C   ammonium carbonate,

D   sodium carbonate or potassium carbonate.

10. An unknown mixture of compounds gave a white precipitate when treated with dilute nitric acid and silver nitrate. The precipitate was filtered off, and a brown ring test was carried out on the filtrate. This test proved positive. The original mixture therefore contained:

A   chloride ions and nitrate ions,

B   chloride ions and possibly nitrate ions,

C   chloride ions only,

D   sulphate ions and nitrate ions.

Chapter 14

Electrochemistry

Electrolysis

Substances which conduct electricity can be divided into two classes.

i. *Metallic conductors*, which conduct electricity without undergoing any chemical change. Examples include all metals, whether solid or molten, and also graphite.

ii. *Electrolytic conductors*, which undergo chemical changes when they conduct electricity. An electrolytic conductor may be a solution of an acid, base or salt in a suitable solvent, which is usually water. Molten, *i.e.* fused, salts and bases are also electrolytic conductors, but *solid* salts and bases are not. Electrolytic conductors are always liquids or solutions, but do not include anhydrous liquid acids, such as pure sulphuric acid.

A compound which conducts electricity (*i.e.* becomes an electrolytic conductor) when dissolved in a suitable solvent or when fused is known as an *electrolyte*. Sodium chloride, for example, is an electrolyte, because when dissolved in water or when molten it conducts electricity. Non-conducting liquids (*e.g.* ethanol), and compounds (*e.g.* sugar) which do not conduct electricity in solution, are termed *non-electrolytes*.

When an electric current is passed through an electrolyte, the electrolyte is decomposed and *electrolysis* is said to take place. A diagram of an apparatus for electrolysis is shown in Fig. 14.1.

Into the electrolyte dip two pieces of a metallic conductor, called *electrodes*. The electrodes should not touch each other and are connected to a direct current supply such as a battery. A 'direct current' consists of a continuous flow of electrons in one direction only, from the negative terminal of the battery to the positive one through the rest of the circuit.

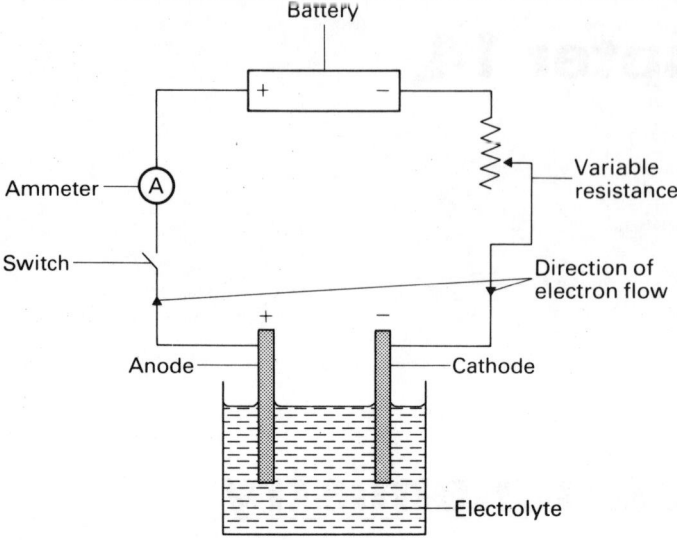

Fig. 14.1 An electrolytic cell or 'voltameter'

The electrode which is connected to the negative battery terminal also carries a negative charge and is called the *cathode*. The other electrode, called the *anode*, carries a positive charge because of its connection to the positive battery terminal. When a direct current is flowing, there is electron movement in two parts of the circuit:

i. from the negative battery terminal to the cathode,
ii. from the anode to the positive battery terminal.

Conduction of electricity through an electrolyte

An electrolyte contains positively charged ions, M^{n+}, and negatively charged ions, X^{n-}. During electrolysis, the positively charged ions are attracted by the negatively charged cathode and move towards it. For this reason we refer to positively charged ions as *cations*. Simultaneously, the negatively charged ions move towards the positively charged anode and are called *anions*.

The conduction of electricity always involves the movement of charged particles, either electrons or ions. Electrolytes conduct electricity by the cations moving to the cathode, and the anions moving towards the anode.

Electrode reactions

When the ions reach their respective electrodes chemical reactions occur.

Reaction at the cathode

The negative charge on the cathode indicates that it carries a surplus of electrons. These can be transferred to the cations of the electrolyte, which thus become converted into atoms. For example, copper(II) ions in solution are *discharged, i.e.* their charge is removed, by gaining electrons from the cathode:

$$Cu^{2+} + 2e^- = Cu$$

Copper is deposited on the cathode, which is said to become *electroplated* with the metal.

Iron(III) ions behave differently in that they are converted into iron(II) ions:

$$Fe^{3+} + e^- = Fe^{2+}$$

However, all ionic half-equations for changes at the cathode involve electron gain, and thus represent reduction reactions. **Always, during electrolysis, reduction occurs at the cathode**; hence we speak of *cathodic reduction*.

Reaction at the anode

The anode carries a positive charge because it has fewer electrons than protons. When anions reach the anode they may give up their electrons to it and become oxidised, as the following ionic half-equations show:

$$2Cl^- = Cl_2 + 2e^-$$
$$4OH^- = 2H_2O + O_2 + 4e^-$$

Alternatively, the anode may dissolve. For example, if a copper anode is used with an electrolyte of sulphuric acid, copper atoms from the anode lose electrons to form copper(II) ions which enter the solution:

$$Cu = Cu^{2+} + 2e^-$$

This, also, is an oxidation process. **During electrolysis, oxidation always occurs at the anode**; we describe this as *anodic oxidation*.

Electrolysis is therefore an oxidation-reduction process.

Some examples of electrolysis

Fused electrolytes with inert electrodes

An inert electrode is one which does not dissolve or become chemically involved in the electrode reactions which occur during electrolysis. Carbon or platinum is frequently used as an inert electrode.

Fused lead(II) bromide When molten lead(II) bromide, $PbBr_2$, is electrolysed, lead is formed at the cathode and bromine is liberated at the anode. Lead(II) bromide contains Pb^{2+} and Br^- ions, and the former are reduced while the latter are oxidised:

$$Pb^{2+} + 2e^- = Pb$$
$$2Br^- = Br_2 + 2e^-$$

The lead may stick to the cathode, which becomes lead plated. Red-brown bromine vapour is evolved from the anode.

Fused sodium chloride Here, sodium ions are discharged at the cathode to produce sodium metal, and chlorine gas is evolved at the anode:

$$Na^+ + e^- = Na$$
$$2Cl^- = Cl_2 + 2e^-$$

Electrolysis of a molten mixture of sodium chloride and calcium chloride is used industrially to manufacture sodium.

Aluminium oxide In the manufacture of aluminium, aluminium oxide, Al_2O_3, is dissolved in a molten salt called 'cryolite', Na_3AlF_6, and the mixture is electrolysed. Aluminium is liberated at the cathode, and oxygen at the anode:

$$Al^{3+} + 3e^- = Al$$
$$2O^{2-} = O_2 + 4e^-$$

Aqueous solutions with inert electrodes

Electrode reactions in aqueous solution are complicated by the fact that more than one type of cation or anion may be present. This is because, in addition to the ions derived from the dissolved compound, there are also a few hydrogen ions and hydroxide ions due to the ionisation of water:

$$H_2O = H^+(aq) + OH^-(aq)$$

The reaction is reversible, and only a very few water molecules are ionised. However, hydrogen ions and hydroxide ions can be discharged to give hydrogen and oxygen respectively:

$$2H^+(aq) + 2e^- = H_2$$
$$4OH^-(aq) = 2H_2O + O_2 + 4e^-$$

If hydrogen ions or hydroxide ions are removed in this way they are replaced, *i.e.* some more water molecules ionise to replace the lost ions. In this way it is possible for hydrogen ions or hydroxide ions to be discharged to the total exclusion of ions from the dissolved substance.

Thus, in aqueous solutions, there is usually a choice of anode reaction and a choice of cathode reaction. We can often predict the outcome because, in general, the ion which is most easily oxidised or reduced at an electrode is discharged in preference to other ions which may be present. The ion is said to be *preferentially discharged*, or *selectively discharged.*

These principles are illustrated by the following examples.

Sodium chloride solution The cations present are sodium ions and hydrogen ions. They are both attracted towards the cathode, and in theory there are two possible reactions that can occur at this electrode:

$$2H^+(aq) + 2e^- = H_2$$
$$Na^+ + e^- = Na$$

Hydrogen ions, however, are preferentially discharged, for they are far more easily reduced than sodium ions. Hydrogen is therefore evolved at the cathode, despite the fact that sodium ions greatly outnumber the hydrogen ions.

The anions present are chloride ions and hydroxide ions. The anode reaction is dependent upon the concentration of the solution. In dilute solution, hydroxide ions are discharged in preference to chloride ions; hence oxygen is formed:

$$4OH^- = 2H_2O + O_2 + 4e^-$$

In concentrated solution, however, chloride ions are selectively discharged and chlorine is formed:

$$2Cl^- = Cl_2 + 2e^-$$

Copper(II) sulphate solution Both copper(II) ions and hydrogen ions are attracted to the cathode. However, the copper(II) ion is selectively discharged and the cathode becomes copper plated:

$$Cu^{2+} + 2e^- = Cu$$

At the anode hydroxide ions are discharged in preference to sulphate ions and bubbles of oxygen are given off. The sulphate ion is extremely difficult to oxidise, and the hydroxide ion is nearly always discharged preferentially.

Solutions of acids and bases Dilute solutions of all common acids (*e.g.* HCl, H_2SO_4 and HNO_3) and all common bases (*e.g.* NaOH and KOH) behave in the same way on electrolysis. In all cases hydrogen ions and hydroxide ions are discharged in preference to other ions in solution, so that hydrogen is evolved at the cathode and oxygen at the anode. Electrolysis is usually carried out in the apparatus shown in Fig. 14.2.

The discharge of hydrogen ions and hydroxide ions means that, in effect, only the water is decomposed. The acid or base in solution remains chemically unchanged, and the entire action can be summarised by the equation:

$$2H_2O = 2H_2 + O_2$$

Experimental evidence is provided by the fact that hydrogen and oxygen are formed in a ratio of almost exactly 2:1, the same as when pure water is used. (Pure water is slow to be decomposed on electrolysis because of its low electrical conductivity.)

Aqueous solutions without inert electrodes
Sodium chloride solution, with a mercury cathode and an inert anode The nature of the cathode can sometimes affect the preferential discharge of cations. This is particularly so in the electrolysis of sodium chloride solution. As stated above, with an inert electrode, hydrogen ions are discharged rather than sodium ions, but with a mercury cathode the order is reversed. The reason is that under these conditions sodium ions are discharged to give *sodium amalgam*, which is a compound of sodium and mercury:

$$Na^+ + e^- + Hg = NaHg$$

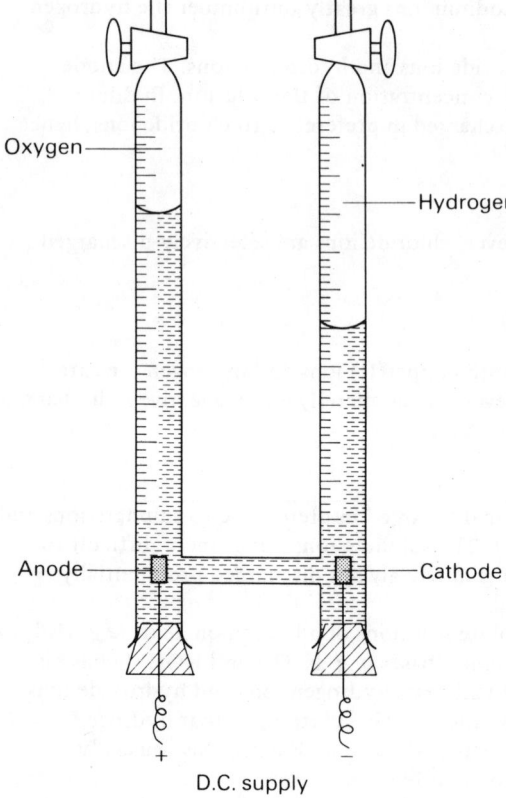

Fig. 14.2 A water voltameter

This is much easier than discharging sodium ions to give pure sodium, and is actually easier than discharging hydrogen ions to give hydrogen at a mercury cathode. If the solution is sufficiently concentrated, chlorine is evolved at the anode.

This forms the basis of a commercial method of making sodium hydroxide. A concentrated solution of sodium chloride is electrolysed with a mercury cathode and a carbon anode. The sodium amalgam formed at the cathode is continuously removed from the cell and allowed to react with water in a separate vessel to form a solution of sodium hydroxide:

$$2NaHg + 2H_2O = 2NaOH + H_2 + 2Hg$$

The mercury produced in this reaction is returned to the cell.

Copper(II) sulphate solution, with a copper anode The nature of the anode can influence the reaction at this electrode, for anions in solution will not be discharged if it is easier for the anode to dissolve. This is illustrated by the

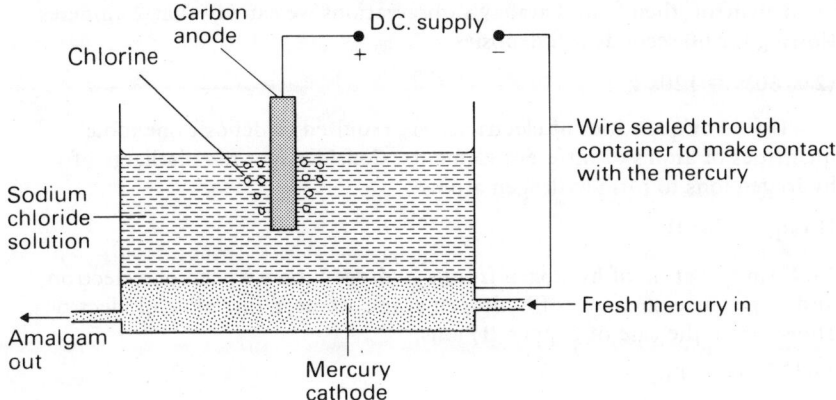

Fig. 14.3 Electrolysis of sodium chloride solution with a mercury cathode

electrolysis of aqueous copper(II) sulphate. Copper(II) ions are always discharged at the cathode, regardless of the material from which the cathode is constructed, but reaction at the anode depends on the nature of this electrode. With an inert anode, as we have seen, hydroxide ions are discharged, but if the anode is made of copper it is easier to oxidise copper atoms, *i.e.*

$$Cu = Cu^{2+} + 2e^-,$$

than it is to oxidise hydroxide ions or sulphate ions. Copper atoms are therefore oxidised preferentially, and enter the solution as hydrated copper(II) ions. The anode thus dissolves, and can be referred to as a *soluble anode*.

To summarise, during the electrolysis of copper sulphate solution with a copper anode, copper(II) ions are discharged at the cathode, but are continually replaced by fresh copper(II) ions from the anode. These principles are utilised in the following industrial processes.

i. Electroplating. An article to be copper plated is made the cathode, while the anode is a bar of pure copper.

ii. Electrolytic purification of copper. Here, the anode is *impure* copper, and the cathode is a thin strip of the pure metal. On electrolysis, impure metal dissolves at the anode and pure metal is deposited on the cathode. Soluble impurities enter the copper sulphate solution, while insoluble impurities form a sludge under the anode.

The Faraday constant

Michael Faraday in 1832 showed that the mass of a substance liberated at an electrode during electrolysis is proportional to the quantity of electricity which passes through the electrolyte.

Quantity of electricity = current flowing × time for which it flows.

Current flowing is measured in *amperes* (A), time in *seconds* (s), and quantity of electricity in *coulombs* (C).

Suppose, for example, that 1 ampere flowing for 1 second deposits x g of an element, then from Faraday's observations we can say that 2 amperes flowing for 60 seconds will deposit:

$(2 \times 60)x = 120x$ g

Different quantities of electricity are required to deposit one mole quantities of atoms of different elements. Consider first the discharge of hydrogen ions to form hydrogen atoms:

$H^+(aq) + e^- = H$

To form one atom of hydrogen from a hydrogen ion we need one electron, and to produce one mole of hydrogen atoms we need one mole of electrons. However, in the case of copper(II) ions,

$Cu^{2+} + 2e^- = Cu,$

two electrons are needed to form one copper atom; thus two moles of electrons are needed to produce one mole of copper atoms.

An electron is a charged particle, *i.e.* it carries a certain quantity of electricity which is measured in coulombs. The quantity of electricity which is carried by one mole of electrons (*i.e.* 6.023×10^{23} electrons) is approximately 96 500 coulombs. This quantity of electricity is called the Faraday constant (F). **The Faraday constant (96 500 C mol^{-1}) is thus the quantity of electricity carried by, or equivalent to, one mole of electrons.**

To deposit one mole of hydrogen atoms we therefore require 96 500 C, *i.e.* F C, and to deposit one mole of copper from copper(II) ions we need $2 \times 96\ 500$ C, *i.e.* $2F$ C. In general, to deposit one mole of an element from its cations, M^{n+}, according to the equation:

$M^{n+} + ne^- = M,$

we require nF C, where n is the number of positive charges carried by the cation. ($n = 1$ for H^+, 2 for Cu^{2+}, 3 for Cr^{3+}, etc.)

To summarise, the quantity of electricity required to deposit one mole of atoms of an element is a whole number of Faraday constants, where the whole number is equal to the valency of the element.

Determination of the quantity of electricity required to deposit one mole of copper

The following instructions relate to the circuit shown in Fig. 14.1, with copper electrodes dipping into copper(II) sulphate solution contained in a beaker.

1. Clean, dry and weigh the copper cathode accurately; then assemble the apparatus.
2. Close the switch, start a stop-clock, and quickly adjust the variable resistance so that a current of approximately 1 ampere flows through the circuit.
3. Make a note of the current flowing, and if necessary adjust the resistance from time to time to maintain the current at this value.

4. Allow the current to flow for about 20 minutes, then switch off, immediately stop the clock, and remove the cathode.
5. Wash the cathode with distilled water and allow it to dry. **Do not wipe the cathode with a cloth, as this may remove some of the deposited copper.**
6. When it is thoroughly dry, reweigh the cathode.
7. If there is time, repeat the procedure by reconnecting the weighed cathode into the circuit.

Calculation Suppose that it is found that 0.337 g of copper is deposited by a current of 0.95 A flowing for 18 minutes, *i.e.* 18 × 60 s.

The quantity of electricity flowing is 0.95 × 18 × 60 = 1026 C.

The relative atomic mass of copper is 63.5; therefore 1 mole of copper is 63.5 g.

If 0.337 g of copper requires 1026 C,

then 63.5 g of copper require $\dfrac{1026}{0.337} \times 63.5 = 193\ 000\ C = 2F$

This result is in accord with the cathode reaction

$$Cu^{2+} + 2e^- = Cu$$

The electrochemical series

The metallic elements vary greatly in their chemical reactivity. When treated with water, for example, sodium and potassium react vigorously in the cold, calcium reacts steadily, magnesium needs to be heated in steam, iron reacts with steam at red heat, and copper and gold do not react at all, even at very high temperatures. If we consider other reactions, such as those with acids, air or oxygen, we find much the same order of reactivity, with sodium and potassium among the most reactive metals and copper and gold among the least reactive.

Whenever a metal undergoes a chemical reaction it becomes oxidised, *i.e.* the metal atoms lose electrons and are converted into positive ions. (This is the reason that metals are defined as electropositive elements.) One of the main factors which determines the reactivity of a metal is the ease with which it can form its ions by electron loss. The most reactive metals are those which readily lose electrons, and the least reactive are those which lose electrons with difficulty.

The ions of highly reactive metals are correspondingly difficult to reduce to the metal, while those of unreactive metals are easily reduced. For example, the reduction of potassium ions to potassium is difficult, but the reduction of copper(II) ions to copper is easily accomplished.

A list of metals can be drawn up based on their relative tendency to form ions in solution. Such a list (see Table 14.1) is known as the *electrochemical series*.

From top to bottom of the electrochemical series metals show a decreasing tendency to form ions. The most reactive metals therefore appear at the

Table 14.1 A selection of metals from the electrochemical series

Potassium	
Calcium	Decrease in electropositive character,
Magnesium	*i.e.* decrease in ease of positive ion formation
Aluminium	
Zinc	Therefore, from top to bottom,
Iron	(i) metals become more difficult to oxidise,
Tin	*i.e.* they become less reactive,
Lead	(ii) metal ions become more easily reduced to
Hydrogen	the metal.
Copper	
Mercury	
Silver	
Gold	

top and the least reactive at the bottom. Notice that hydrogen is also included in the list, despite the fact that it is not a metal. Hydrogen is, however, an electropositive element, because it forms positive ions.

Because they form positive ions relatively easily, the elements at the top are said to be 'more electropositive' than those lower down. Thus, **there is a decrease in electropositive character from top to bottom of the electrochemical series.**

Deduction of the electrochemical series from displacement reactions

In general, metal A will displace metal B from a solution of one of its salts provided that metal A is higher in the electrochemical series than metal B. Zinc, for example, is higher in the electrochemical series than copper, and when a piece of zinc is immersed in a solution of copper(II) sulphate it becomes coated with copper. A chemical reaction occurs between the zinc and the copper(II) sulphate, as a result of which some of the zinc enters solution as zinc sulphate, while some of the copper(II) sulphate is converted into copper:

$$Zn + CuSO_4 = ZnSO_4 + Cu$$

We say that 'zinc displaces copper from copper(II) sulphate solution' in a *displacement reaction*. Like all other instances of displacement, this is essentially a redox reaction. Copper(II) sulphate and zinc sulphate are electrovalent compounds, *i.e.* $Cu^{2+} SO_4^{2-}$ and $Zn^{2+} SO_4^{2-}$ respectively. Sulphate ions are spectator ions, so that the ionic equation is as follows:

$$Zn + Cu^{2+} = Zn^{2+} + Cu$$

The displacement reaction therefore involves the oxidation of zinc,

$$Zn = Zn^{2+} + 2e^-,$$

and the reduction of copper(II) ions by the electrons that are lost from the zinc:

$$Cu^{2+} + 2e^- = Cu$$

Electrons are transferred and the reaction occurs because zinc has a greater tendency than copper to exist in the form of ions.

We can attempt other displacement reactions with other metals and salts and, by similar reasoning, construct in part the series shown in Table 14.1. For example, we can show that zinc displaces not only copper but also tin and lead from solutions of their salts:

$$Zn + Pb(NO_3)_2 = Zn(NO_3)_2 + Pb$$
$$Zn + SnCl_2 = ZnCl_2 + Sn$$

Zinc is therefore above all three metals in the electrochemical series. Tin is above lead and copper, because it displaces both these metals, and lead displaces copper. The relative positions of these elements are thus:

zinc
tin
lead
copper

The reasoning can be applied also to hydrogen, in that metals above hydrogen in the series will displace hydrogen from dilute solutions of acids (except nitric acid), but metals below hydrogen will not do so.

Uses of the electrochemical series

Reactivity guide

The electrochemical series provides a good guide to the reactivity of metals towards such reagents as water, acids or air. Metals become decreasingly reactive towards these reagents from top to bottom of the series, as it becomes increasingly difficult to convert them to their ions. The reactions with water and acids have been discussed previously, and we shall now consider those with air.

Metals at the top of the series, such as potassium and sodium, oxidise rapidly on contact with the air and for this reason must be stored under liquid paraffin. Calcium reacts less readily than potassium or sodium and need not be kept immersed in liquid paraffin. The reaction product in all three cases is initially the oxide of the metal, which then reacts with water vapour and carbon dioxide in the atmosphere to form the carbonate, e.g.

$$4Na + O_2 = 2Na_2O$$
$$Na_2O + H_2O = 2NaOH$$
$$2NaOH + CO_2 = Na_2CO_3 + H_2O$$

Magnesium and aluminium initially react very rapidly with air to give their oxides. In both cases, however, the oxide forms a thin, imperceptible coating on the metal, and this prevents further attack by the air. Metals below aluminium in the series either react very slowly or do not react at all with dry air at room temperature.

Certain metals, notably iron, are attacked by air far more rapidly in the wet than in the dry. Both water and carbon dioxide, as well as oxygen, contribute to the rusting of iron.

The selection of methods for the reduction of metal ores

The ease with which a metal ore is reduced to a metal depends on the position of the metal in the electrochemical series. Ores of metals at the top of the series are the most difficult to reduce, but on descending the series the reduction of ores becomes progressively easier.

The ores of metals at the bottom of the series are reduced very easily indeed. Some metals, notably gold, are so unreactive that they are found *native, i.e.* in the elemental state, and a simple physical separation from earth and rocks is all that is required for their isolation.

The common elements below aluminium, *i.e.* iron, zinc, lead, copper and tin, are obtained by reducing the ores or oxides with carbon or carbon monoxide.

For elements above and including aluminium, it is difficult to find reducing agents which are both suitable and convenient. The ions of these elements are difficult to reduce, and very powerful reducing agents are needed. Such elements are always produced by the electrolysis of their compounds, usually fused salts under anhydrous conditions (see the first part of this chapter). Electrolysis is a simple means of achieving both reduction and oxidation changes which are difficult to bring about by chemical reaction.

Summary

At the conclusion of this chapter, you should be able to:
1. distinguish between metallic and electrolytic conductors,
2. recognise that electrolytes conduct electricity by ionic movement,
3. recognise that electrolysis is accompanied by reactions at the electrodes,
4. recognise that oxidation occurs at the anode, *i.e.* the positive electrode, and reduction at the cathode, *i.e.* the negative electrode,
5. represent electrode reactions by ionic half-equations,
6. describe the electrolysis of fused salts,
7. describe the electrolysis of aqueous sodium chloride with inert electrodes in terms of the selective discharge of ions,
8. describe the effects of change of electrode material in the electrolysis of aqueous sodium chloride or copper(II) sulphate,
9. define 'quantity of electricity' as current flowing X time, and state its units as coulombs,
10. state that the mass of a substance liberated on electrolysis is proportional to the quantity of electricity which is used,
11. define the 'Faraday constant' and state its value,
12. state that the quantity of electricity required in electrolysis to deposit one mole of atoms of an element is a whole number of Faraday constants,

13. determine the quantity of electricity necessary to deposit one mole of copper,
14. recognise that the electrochemical series is based on the relative tendency of metals to form ions in solution,
15. describe how the electrochemical series can be deduced in part from displacement reactions,
16. relate the electrochemical series to the trends observed in the reactions of metals with water, acids and air,
17. predict the reactivity of a metal, and the reducibility of its ore, from the position of the metal in the electrochemical series,
18. select a method for the extraction of a metal from its position in the electrochemical series.

Questions

1. Which of the following statements about electrolysis are true and which are false?
   (a) Negatively charged ions are called cations.
   (b) The positively charged electrode is called the anode.
   (c) A direct current supply must be used.
   (d) Positively charged ions travel towards the cathode.
   (e) Solid sodium chloride is an electrolytic conductor.
   (f) Pure sulphuric acid is not an electrolytic conductor.
   (g) Electrons flow from the positive battery terminal to the anode.
   (h) Reaction at the cathode always involves the liberation of electrons.
   In questions 2—6 select the most appropriate answer, labelled A, B, C or D.

2. In the electrolysis of aqueous copper(II) sulphate with inert electrodes,
   A   copper is deposited on the anode,
   B   one mole of electrons is required to deposit one mole of copper,
   C   sulphate ions are discharged at the anode,
   D   oxygen gas is evolved at the anode.

3. In the electrolysis of concentrated aqueous sodium chloride with a mercury cathode,
   A   hydrogen ions are selectively discharged at the cathode,
   B   oxygen is evolved at the anode,
   C   sodium ions are discharged in preference to hydrogen ions,
   D   sodium hydroxide is formed at the anode.

4. If one coulomb of electricity deposits 3.37×10^{-4} g of zinc on electrolysis, what mass of zinc is deposited by a current of 2 amperes flowing for 2 minutes?
   A 6.74×10^{-4} g
   B 1.348×10^{-3} g
   C 4.044×10^{-2} g
   D 8.088×10^{-2} g

5. Which one of the following statements concerning the Faraday constant is *untrue*?

A   It has a value of approximately 96 500 C mol^{-1}.

B   It is equivalent to one mole of electrons.

C   It will liberate one mole of molecular hydrogen, H_2 (g), on the electrolysis of dilute sulphuric acid.

D   It will liberate half a mole of copper on the electrolysis of aqueous copper(II) sulphate.

6. Which one of the following properties increases from top to bottom of the electrochemical series?

A   Electropositive character.

B   Ease of oxidation.

C   Reactivity towards air.

D   Ease of reduction of metal ions.

For each of the metals shown in questions 7—10, select the most suitable reduction method from the list below for the extraction of the metal from its ore.

A   Electrolysis of the chloride

B   Electrolysis of the oxide

C   Heating the oxide with carbon

7. Iron

8. Tin

9. Sodium

10. Aluminium

Answers

Chapter 1 (page 7)
3, 5, 7 and **8** are chemical changes, and **1, 2, 4** and **6** are physical changes.

Chapter 2 (page 18)
1.C 2.B 3.C 4.A 5.D
6. (a) (d) (e) and (i) are elements, (b) (f) and (g) are compounds, and (c) (h) and (j) are mixtures.

Chapter 3 (page 26)
1. (a) 16 (b) 16 (c) 16 (d) 2 (e) 8 (f) 6 (g) 32 (h) +16
2. (a) and (h) are true, and (b) (c) (d) (e) (f) and (g) are false.
3. (a) (e) (g) and (h) are permissible, and (b) (c) (d) and (f) are not.
4.A 5.B 6.C

Chapter 4 (page 42)
1. (b) (d) and (f) are metals, and (a) (c) (e) (g) and (h) are non-metals.
2. (a) Ar (b) Al (c) Na (d) Sr (e) Sn (f) Ti (g) Mg (h) Mn
3. (a) Potassium (b) Lead (c) Silver (d) Arsenic (e) Gold (f) Iron (g) Chromium (h) Mercury
4. (a) Cl_2 (b) O_2 (c) N_2 (d) P_4 (e) C
5. (b) (c) (f) and (h) exhibit variable valency.
6.B 7.D 8.B 9.A 10.D 11.D 12.C 13.D

Chapter 5 (page 64)
1.D 2.B 3.C 4.A 5.D 6.C 7.C 8.D 9.B 10.C
11. (a) $MgCO_3$ (b) $ZnBr_2$ (c) K_3PO_4 (d) $Al(OH)_3$ (e) FeO (f) Fe_2O_3 (g) NH_4NO_3 (h) Na_2S (i) Na_2SO_3 (j) $(CH_3COO)_2Ca$ or $Ca(OCOCH_3)_2$
12. (a) Aluminium chloride (b) Potassium carbonate (c) Sodium nitrite (d) Sodium nitrate (e) Copper(II) oxide (f) Copper(I) oxide (g) Potassium hydrogensulphate (h) Sodium hypochlorite (i) Calcium hydrogencarbonate (j) Chromium(III) chloride
13. (b) and (e) are acids, (c) (d) and (h) are bases, and (a) (f) (g) (i) and (j) are salts.

Chapter 6 (page 75)
1. (a) 63 (b) 106 (c) 2.0% of hydrogen, 32.7% of sulphur and 65.3% of oxygen (d) 62.9% (e) 56 g (f) 171 g (g) 1.71 mol (h) 0.004 mol
2. (b) (c) (e) and (g) are true, and (a) (d) (f) and (h) are false.

Chapter 7 (page 91)
1. (*a*) (*b*) (*g*) and (*j*) are true, and (*c*) (*d*) (*e*) (*f*) (*h*) and (*i*) are false.
2. 10 g **3.** 1170 g **4.** 0.53 g **5.** 0.822 mol dm^{-3} **6.** 0.602 mol dm^{-3}
    7. 0.064 mol dm^{-3} **8.** 0.059 mol dm^{-3} **9.** 4.9 g dm^{-3} **10.** 5.6 g dm^{-3}
    11. 298 g dm^{-3} **12.** B **13.** D **14.** D **15.** B **16.** A **17.** D **18.** C
19. (*a*) 53 g per 100 g, (*b*) 130 g per 100 g.

Chapter 8 (page 104)
1. (*b*) (*d*) (*g*) and (*h*) are molecular crystals, (*c*) (*e*) and (*j*) are atomic
    crystals, (*i*) is an ionic crystal, and (*a*) and (*f*) are metallic crystals.
2. A,D,G **3.** B,D,E **4.** B,C,F **5.** B,D,E **6.** A,D,G **7.** B **8.** B **9.** C **10.** A
11. A **12.** C **13.** D

Chapter 9 (page 127)
1. (*a*) $CuCO_3 = CuO + CO_2$
    (*b*) $2NaHCO_3 = Na_2CO_3 + CO_2 + H_2O$
    (*c*) $2Fe + 3Cl_2 = 2FeCl_3$
    (*d*) $4Al + 3O_2 = 2Al_2O_3$
    (*e*) $Ca + 2H_2O = Ca(OH)_2 + H_2$
    (*f*) $2Al + 3H_2SO_4 = Al_2(SO_4)_3 + 3H_2$
2. 1.38 g **3.** 15.3 g **4.** (*a*) 161.0 g (*b*) 147.9 g **5.** (*a*) 3.40 g (*b*) 4.48 dm^3
    (*c*) 5.06 dm^3 **6.** 6.43 g **7.** 1.90 g **8.** 1 mol **9.** B **10.** C **11.** C **12.** B
13. D

Chapter 10 (page 138)
1. (*a*) (*c*) (*d*) (*f*) and (*g*) are basic, and (*b*) (*e*) and (*h*) are acidic.
2. (*a*) Fe_3O_4, basic (*b*) CaO, basic (*c*) H_2O, neutral (*d*) SO_2, acidic
    (*e*) P_4O_{10}, acidic (or P_4O_6) (*f*) CuO, basic (*g*) CO_2, acidic (*h*) Na_2O,
    basic (or Na_2O_2)
3. C **4.** A **5.** D **6.** A **7.** D **8.** A

Chapter 11 (page 147)
1. (*a*) H^+ (hydrogen ion) + Cl^- (chloride ion), strong
    (*b*) H^+ + SO_4^{2-} (sulphate ion), strong (*c*) H^+ + NO_3^- (nitrate ion), strong
    (*d*) H^+ + CH_3COO^- (ethanoate or acetate ion), weak
    (*e*) Na^+ (sodium ion) + OH^- (hydroxide ion), strong
    (*f*) NH_4^+ (ammonium ion) + OH^-, weak
2. C **3.** B **4.** D **5.** A **6.** B **7.** D
8. (*d*) (*g*) and (*h*) are true, and (*a*) (*b*) (*c*) (*e*) and (*f*) are false.

Chapter 12 (page 171)
1. C **2.** C **3.** A **4.** B **5.** D **6.** B
7. (*a*) 10.0 cm^3 (*b*) 26.25 cm^3 (*c*) 23.8 cm^3 (*d*) 200.0 cm^3 (*e*) 158.7 cm^3
    (*f*) 23.8 cm^3

8. (a) 0.0491 mol dm^{-3} (b) 0.0915 mol dm^{-3} (c) 0.0833 mol dm^{-3}
    (d) 3.33 g dm^{-3}
9. A 10. C 11. B 12. A 13. A
14. (b) (c) and (k) are true, and (a) (d) (e) (f) (g) (h) (i) and (j) are false.
15. F,G,H 16. B 17. B 18. B 19. A 20. B 21. A 22. C 23. B 24. A

Chapter 13 (page 183)
1. C 2. B 3. D 4. A 5. A 6. C
7. (a) and (c) are true, and (b) and (d) are false.
8. A 9. D 10. A

Chapter 14 (page 197)
1. (b) (c) (d) and (f) are true, and (a) (e) (g) and (h) are false.
2. D 3. C 4. D 5. C 6. D 7. C 8. C 9. A 10. B

Index